SYMBOLS AND SACRIFICE IN WAR

SYMBOLS AND SACRIFICE IN WAR

National Identity and the Will to Fight

KIRSTIN J.H. BRATHWAITE

GEORGETOWN UNIVERSITY PRESS / WASHINGTON, DC

© 2025 Georgetown University Press. All rights reserved. No part of this book may be reproduced or utilized in any form or by any means, electronic or mechanical, including photocopying and recording, or by any information storage and retrieval system, without permission in writing from the publisher.

The publisher is not responsible for third-party websites or their content. URL links were active at time of publication.

Library of Congress Cataloging-in-Publication Data

Names: Brathwaite, Kirstin J. H., author.
Title: Symbols and sacrifice in war: national identity and the will to fight / Kirstin J. H. Brathwaite.
Description: Washington, DC: Georgetown University Press, [2025] | Include bibliographical references and index.
Identifiers: LCCN 2024025651 (print) | LCCN 2024025652 (ebook) | ISBN 9781647125783 (hardcover) | ISBN 9781647125790 (paperback) | ISBN 9781647125806 (ebook)
Subjects: LCSH: Military morale. | Operational readiness (Military science) | Nationalism. | Psychology, Military. | World War, 1939-1945—Psychological aspects. | Soldiers—Psychology.
Classification: LCC U22 .B714 2025 (print) | LCC U22 (ebook) | DDC 355.1/23—dc23/eng/20241015

LC record available at https://lccn.loc.gov/2024025651
LC ebook record available at https://lccn.loc.gov/2024025652

∞ This paper meets the requirements of ANSI/NISO Z39.48-1992 (Permanence of Paper).

EU GPSR Authorized Representative
LOGOS EUROPE, 9 rue Nicolas Poussin,
17000, LA ROCHELLE, France
Email: Contact@logoseurope.eu

26 25   9 8 7 6 5 4 3 2   First printing

Printed in the United States of America

Cover design by Jim Keller
Cover photograph of Allied soldiers at Tobruk, Libya, October 1941, from the Australian War Memorial, Order #18695885.
Interior design by Westchester Publishing Services

To John and Mary Hasler—thank you for teaching us to love books.

CONTENTS

List of Illustrations   ix

Acknowledgments   xi

Introduction   xiii

1   Explaining Will to Fight   1

2   National Identity, Democracy, and Cohesion in the British Imperial Forces   28

3   North Africa   67

4   Malaya   116

5   Europe   149

Conclusion   191

Bibliography   203

Index   217

About the Author   221

ILLUSTRATIONS

TABLES

1.1   Predicted Interactions Between Identity and War Goals   3
1.2   Actual Levels of Will to Fight   19
C.1   Theoretical Explanations for Levels of Will to Fight   192

ACKNOWLEDGMENTS

This book has been a long time in the making, which means I have quite a few people to thank for their help and support along the way. The operationalization of will to fight as well as some of the Malaya chapter were originally published in "Effective in Battle: Conceptualizing Soldiers' Combat Effectiveness," *Defense Studies* (Vol. 18, No. 1, 2018).[1] Thank you to Taylor and Francis for permission to reproduce that material here. The Nanovic Center for European Studies and the Institute for Scholarship in the Liberal Arts at the University of Notre Dame both provided funding for early archival work. James Madison College at Michigan State University also funded a later archival trip as well as invaluable work from a number of amazing research assistants. Noelle Cohn, Billy Gischia, Brenna Wayne, Sydney Lipton, Jade McGuffey, Lauryn Purches, and Paige Rissman all spent hours transcribing archival documents from my fuzzy photos into usable Word documents, helping me comb through those documents, and helping to check my references. Erin Hasler read the final draft with great care to deal with my horrible spelling. Without their work this book would have taken even longer (if that is possible). I thank each of you for your hard work and for the chance to collaborate with you.

Several amazing colleagues read drafts along the way, giving invaluable feedback. (The remaining flaws are my own.) A number of ISA panelists and participants in the University of Virginia National Security Policy Center asked insightful questions and provided excellent suggestions. Michael Desch, Charles Fagan, Margarita Konaev, Jeffrey Meiser, Dan Philpott, Michael Posnansky, Sebastian Rosato, Laura Taylor, and Will Walldorf all read at least part (and generally more) of the manuscript. I want to especially thank Will, who not only provided invaluable feedback and advice throughout the writing process but also introduced me to the study of international relations as an undergraduate. He has been a generous mentor for many years, and my career would not be the same without him. Other Gordon College faculty were also instrumental in my early training in political science and my development as a scholar. Stephen Alter, Agnes Howard, Tal Howard, Ruth Melkonian Hoover, and Timothy Sherratt have my deepest thanks for their encouragement, teaching, and examples.

Though the writing process can seem a solitary one, I have benefitted from an amazing community of scholars at Michigan State University. Matthew Zierler and Mark Axelrod were invaluable mentors in my first few years as a professor and have continued to provide much-needed support and advice, as have my deans: Sherman Garnett, Brian Johnson, Jeffrey Judge, Linda Racioppi, and Cameron Thies. Ani Sarkissian and Mickey Stam have been great friends and mentors since I first arrived at MSU. Omowumi Olufunmbi Elemo-Kaka and I traveled the fixed-term to tenure-track road together; I don't know how I would have managed without her. Angelina Sandora has been a voice of encouragement and support. And Susan Stein-Roggenbuck has been a sounding board, mentor, guide, and friend from day one. Thank you for everything you do, Susan.

My family has been encouraging, motivating, and supportive of my academic pursuits since I was a child. Thanks, Mom and Dad, for the sacrifices you made to make sure we all had the best education you could imagine. My in-laws, my siblings and their partners are a wonderful support network: Lannie, Billie and Burt, Anna, Josh and Laura, Sarah and Sam, Rachel and Josh, and Erin—thanks for being awesome and for only making fun of me some of the time. And mostly thanks for all my amazing nieces and nephews. Kauai, Riley, Avery, Benjamin, Maisey, Willow, and Samuel: You are so much fun and I'm so glad to be your aunt!

Finally, I cannot really explain what it has meant to have my husband, Robert Brathwaite's, support and love throughout this process. He gave advice, talked me off the ledge, provided ice cream, and was just generally the best partner imaginable. Thank you.

NOTE

1. Tandfonline.com

INTRODUCTION

In the winter of 1939, the Finnish Army held off the massive Red Army invasion of Finland for three months. With Stalin determined to demonstrate the might of his rebuilt forces and acquire territory, Finland was unlikely to prevail. Yet its army stayed in the battle, fighting with determination and creativity until overwhelmed by Soviet numbers.[1] Why were Finnish soldiers willing to risk their lives in this doomed fight?

Faced with a much weaker opponent, the American military in Vietnam, on the other hand, had the numbers and combat power the Finns could only dream of. Yet after the Tet Offensive in 1968 (which the Americans incidentally won), military morale and discipline declined precipitously. American soldiers were frustrated and disillusioned with their situation. As discipline began to break down, drug abuse increased and attacks on officers by their own units grew increasingly common.[2] What explains this decline in the American military's willingness to keep fighting?

More recently, Iraqi army units (equipped and trained by the US military) collapsed when confronted by Islamic State of Iraq and Syria (ISIS) units invading from Syria. The Iraqi units disintegrated, failing to defend their own territory from lightly armed rebel invaders. Why were ISIS fighters willing to challenge a materially superior force in its own territory while that force was unwilling to defend that territory?

Each of these examples highlights an important question: Why are some militaries more willing to fight than others? There is a great deal of scholarship in international security dedicated to answering this question, much of it focused on calculating the balance of material resources that military organizations have at their disposal and understanding the strategies they adopt to make use of those resources. A growing strand of research, however, also looks to the human resources available to militaries: the soldiers (or potential soldiers) who must put material resources and technology to work in any war.[3]

This book focuses on this human element of combat, asking why some soldiers seem more willing than others to risk and even sacrifice their lives in battle, often when faced with unlikely odds.[4] In what I call national identity theory, I argue that when a military holds to a common national identity, and when soldiers view the goals of the war as compatible with their national identity, that military will have high will to fight. National identity theory builds on three existing explanations for will to fight: small-unit cohesion, political structure, and threat. To develop the argument, I draw on a growing literature that emphasizes the importance of emotion, myth, and narrative in international politics to explain how national identity and the cause of the war combine to influence soldiers' will to fight.

That nationalism, identity, and politics motivate soldiers at times of conflict is not a new insight. My contribution is to develop a theory explaining the *process* by which national identity and politics lead to motivation on the battlefield, as well as the conditions under which they do so. Simply having a common identity will not necessarily be enough to motivate soldiers. Members of that identity group have to believe that the goals of the war—the purpose for which they fight—are in line with the core myths and narratives of their identity in order to be willing to fight. To evaluate the argument's plausibility, I adopt a qualitative research design using comparative case studies of British, Australian, and Indian units fighting in WWII. Because these different nationalities served together in many of the same campaigns, I can hold constant opponent, material circumstances, equipment, and doctrine while varying national identity. I draw on extensive archival resources as well as secondary historical work to explore the motivation of these units across battles in Europe, North Africa, and Malaya.

The historical case research in this book speaks to a modern problem, and not one that is purely intellectual in nature. Whether it is the Kurdish militias fending off ISIS in northern Syria or the rapid collapse of the Afghan army after the ignominious US withdrawal from Kabul in the summer of 2021, the outcomes of battle are determined not simply by firepower and numbers but by the willingness of soldiers to stay in the fight or take the fight to the enemy. Indeed, political and military leaders making decisions about anything from the use of force to manpower policy have intrinsically held beliefs about combat motivations.[5] For example, arguing that unit cohesion motivates soldiers, US officials have at various times excluded African Americans, women, and LGBTQ+ individuals from joining or fully integrating into the military for fear of disrupting that cohesion.[6] Understanding what drives combat motivation is an inquiry into the very essence of war, politics, and the ties that bind soldiers together during humanity's darkest hours.

WHAT IS WILL TO FIGHT?

How well soldiers perform in battle is determined by their skills and material resources, but also by having the will to put those skills and materials to use in a fight. In other words, vast military supplies, advanced technology, and brilliant strategy all require combatants willing to put them to work in battle. As such, will to fight is a key element of combat effectiveness and is therefore pertinent to explaining victory and defeat.[7] I describe the concept of will below, while the next chapter addresses the different measures and indicators used to operationalize this multifaceted notion.

A recent RAND report explains will to fight as "the disposition and decision to fight, to act, or to persevere when needed."[8] Simply put, will to fight is a soldier's willingness to engage in battle.[9] Scholars typically discuss three components that make up will to fight: the level of soldiers' morale, discipline, and initiative in battle.[10]

Soldiers with high morale, who are enthusiastic and persistent in their commitment to victory are less likely to desert, to shirk duties, or fail to fire their weapons.[11] Morale is notoriously difficult to define and on its own cannot explain willingness to fight. Yet soldiers' attitudes or thinking regarding their situation certainly influences how they behave on the battlefield and remains an important aspect of motivation to fight.[12]

Discipline is the defining feature of professional military training and is widely recognized as fundamental to successful performance in battle.[13] Some of this has to do with the fact that soldiers who follow orders without threat of the use of force reduce the manpower and material resources devoted to policing a military's own troops. Yet, as Jason Lyall notes, soldiers sometimes persevere in battle not out of willingness to fight but due to threats from their own forces—blocking, barriers, or anti-retreat formations.[14] The need to put military resources into forcing soldiers to remain on the battlefield suggests that discipline has already broken down and soldiers are not so much willing to engage in battle as unwilling to face death from their own compatriots. The collapse of discipline is oftentimes a harbinger of desertion and defeat.

Finally, soldiers who have initiative, who are willing to seek out engagement with the enemy and try new tactics, take a proactive approach to battle and create opportunities.[15] Initiative requires them to take on additional risks, moving beyond their training in order to maintain engagement with the enemy. It is critical to note that under this conceptualization, initiative and discipline are not at odds. Initiative is not ignoring orders but finding the best way to carry them out. Martin Van Creveld has argued that one reason German soldiers were so successful in battle was that officers expected them to show initiative and creativity in accomplishing their assigned tasks or goals.[16] Similarly, Stephen Ambrose argues that the American military was successful in WWII in part due to the creativity of the soldiers who figured out how to overcome unexpected obstacles.[17]

Will to fight thus encompasses both attitude toward the fight and behavior in battle. Morale, discipline, and initiative in battle all combine into soldiers who are willing to put the material resources and training they receive to use in combat. Will alone cannot explain victory or defeat, but it is a vital aspect of overall combat effectiveness. A military without soldiers willing to fight will be hard-pressed to take advantage of material or doctrinal resources.

WHAT MOTIVATES SOLDIERS IN BATTLE?

Why soldiers risk and sometimes sacrifice their lives in battle, and why they are willing to take the lives of others, is a fundamental question in both the study and practice of war. Scholarship addressing it ranges from psychological studies of individual propensities to sociological or political studies of the tendencies of states.[18] Human motivation is a complex phenomenon, and no single factor can fully explain the willingness of people to risk their lives and take the lives of others in battle. A unitary theory cannot explain all aspects of soldiers' willingness to engage in battle.[19] The extensive literature dedicated to understanding will to fight bears out the complexity of the issue.

The most well-known explanation for will to fight is small-unit or primary-group cohesion—the "band of brothers" argument. According to this argument, soldiers are generally apolitical and are most concerned with surviving battle.[20] The best way to ensure survival is to build trust and cooperation with others in one's unit.[21] Soldiers are then more willing to fight because they have confidence in their unit and are motivated to ensure its survival. Additionally, the unit provides soldiers with a sense of belonging and emotional support under extremely stressful circumstances.[22] A related argument focuses on secondary cohesion, loyalty to the regiment/battalion/organization beyond the combat platoon.[23] Focus on esprit de corps, or commitment to regimental honor and history, is designed to develop loyalty to the larger military organization and its task of achieving victory in battle.[24]

Another set of arguments points to the political structure of the society from which the soldiers are drawn. Scholarship on the democratic peace suggests that democracies are more effective in battle than non-democracies.[25] There are several pathways through which democracy influences battlefield performance. First, democracies tend to be wealthier and have more loyal soldiers who perform better in battle.[26] Similarly, states where political discourse is free and the rights of individuals are protected are more likely to produce soldiers who are self-motivated and creative.[27] Democracy thus leads especially to initiative in battle, a key element of will to fight. Moreover, democracies tend to select into wars that they are likely to win. Though this selection effect tends to focus on material capability, it may also be the case that democratic governments consider their citizens' likely motivations in potential wars as part of their calculus.[28]

A subset of arguments draws on aspects of both the cohesion and democracy theories to explain why soldiers quit, in particular why they surrender. These authors point to pragmatism and battlefield experience as key in understanding surrenders.[29] Soldiers are rational actors weighing survival versus liberty in deciding whether to fight or surrender. Ryan Grauer argues that soldiers' expectations regarding how they will be treated during captivity and the length of that captivity influence their choices. Soldiers who believe they will be treated decently and their captivity will be short are more likely to surrender.[30] There are a number of possible ways that soldiers make these determinations. Reiter and Stam argue that democracies treat prisoners better and so soldiers are more likely to surrender to democratic opponents.[31] James Morrow argues that battlefield rumors of mistreatment affect both sides, as soldiers expect reciprocity. If their own side treats prisoners badly, they will expect to be treated badly themselves, thus reducing their likelihood of surrender.[32] Grauer points to conditions at home for soldiers who were imprisoned as well as the momentum of the war as indicators soldiers use to determine the risks of surrender, while Todd Lehmann and Yuri Zhukov argue that soldiers are more likely to surrender as units, and do so based on precedents from recent battles.[33]

Finally, war is about threat—threat to the state or the independence of the political community. The argument here is that the more direct and pressing the threat to the state, the more motivated its soldiers will be to fight.[34] Realist scholars label this effect *nationalism*. In this telling, soldiers identify the nation with the state and are motivated to defend their nation-state from external threats. The survival of the nation depends on the survival of the state, and thus the more powerful and imminent the threat to the state, the more motivated soldiers will be to defend against it. Notably, nationalism is really epiphenomenal to threat in this argument. The real work here is done by the level of threat posed to the state.

Taken together, each of these arguments explains important elements of will to fight, but each leaves key questions unanswered as well. While primary-group cohesion is certainly cited by soldiers as a source of motivation, it cannot explain continued motivation in situations of high casualties where soldiers do not know the people with whom they serve nor feel a commitment to the unit's honor, as was the case for the Wehrmacht on the Eastern Front in WWII.[35] Additionally, if soldiers are most motivated by close relationships with members of their unit and a desire to have the unit survive, then avoiding battle is just as reasonable—if not more so, depending on the situation—as fighting hard.[36] What keeps the cohesive unit committed to the fight rather than looking for a way to save the unit from potential destruction in the fight?

Though education, individual resourcefulness, and creativity certainly contribute to initiative, the democratic effectiveness literature has not found much support for commitment to democratic government improving soldiers' morale.[37] There is significant evidence that democratic leaders and their militaries face political pressures regarding the progress of the war that may influence decisions about leadership and strategy.[38] Yet democracies are clearly more sensitive to their citizens' willingness to bear the costs of war than non-democracies. It may be that considering democratic politics in conjunction with national identity and the goals of the war will reveal new patterns of combat motivation. Scholarship about surrender rightly points to soldiers' pragmatism but doesn't account for the value soldiers may place on the outcome of the war or the importance of emotion in such stressful circumstances. Threat, too, suffers from a similar problem in explaining variation across and within cases as the threat posed by an opponent may become more or less dire over the course of a conflict. It also assumes that soldiers identify their survival with their state's survival, which may not be the case; consider the fact that people join separatist rebellions that seek to break up the state. To understand what soldiers perceive as threatening, we must understand the communities with which they identify and may seek to protect.

For a more comprehensive assessment, a number of scholars have argued that some kind of an ideological or identity-based motivation is necessary to complete the explanation of will to fight. While studying American soldiers in the Vietnam War, Charles Moskos notes the importance of small-unit cohesion but also suggests an additional necessary factor—what he calls *latent ideology*. Defining latent ideology as a belief in the worthwhileness of American society, he argues that soldiers seemed to have lost this belief toward the end of the war, thus explaining some of the breakdown in combat motivation in the final years of that war.[39] Darryl Henderson also notes the importance of "internalization of values" for motivating soldiers in the North Vietnamese and Israeli armies, while the lack of such values explains a lack of motivation among American soldiers in Vietnam.[40] H. Wayne Moyer argues that soldiers are motivated in part by a national consensus supporting the cause of the war they are fighting; he focuses on political debate in Congress as indicative of the level of social support soldiers may feel they have.[41] Anthony King points out the ways states mobilized gender norms and political ideology to motivate soldiers throughout the twentieth and twenty-first centuries.[42] Jasen Castillo argues that the degree of ideological commitment in society, combined with the ability of the military to properly train its members, influences soldiers' loyalty on the battlefield.[43] Omer Bartov argues for the importance of the ideological cause to Wehrmacht soldiers on the Eastern Front, while Stephen Fritz also suggests that soldiers viewed National Socialism as having restored German strength and identity.[44] Strong ideological cohesion leads to motivation in battle. Building on these insights, this book offers a theory of how identity and political cause work together to motivate soldiers in battle—and when they do not.

NATIONAL IDENTITY AND WILL TO FIGHT

National identity theory as an explanation for combat motivation is rooted in both the realist insight that threat to the political community fuels soldiers' will to fight and the growing literature on the role of emotion, myth, and narrative in international politics. The central argument is that when leaders are able to convincingly frame the goals of the war in terms of nationally significant myths and symbols, it legitimizes the war and sparks group emotions that motivate soldiers to fight. But what is national identity, and how is it connected to will to fight?

Nationalism and National Identity

Nationalism is commonly connected to combat motivation, and soldiers from nationalist states are often portrayed as more loyal, more aggressive, and more committed to winning the fight.[45] Yet history and historical research show mixed evidence about the precise nature of this relationship.[46] Vichy leaders in WWII France, for instance, claimed nationalism as their motivator in collaborating with Germany—the best way to ensure the survival of the nation was to end the fight.[47] When the subnations of Yugoslavia began to break away from the rump state, there was significant desertion from the Yugoslav state army (JNA) as soldiers chose ethno-national identity over loyalty to the state.[48] Still, there is also strong evidence that nationalist appeals do sometimes motivate combatants. John Lynn details the ways French peasants were motivated by a newly formed identification with the French nation during the Napoleonic wars, while Roger Reese argues that Soviet soldiers fought in part out of love for country, even if they bore little love for the Stalinist regime.[49]

This book strives to reconcile these conflicting accounts, acknowledging the power of nationalism in fueling combat motivation but also offering a nuanced account of the process by which this power takes hold and the circumstances under which it operates. I argue that the role of nationalism in producing will to fight varies depending on the relationship between the goals of the war and the myths and symbols that make up the content of nationalism—what I call *national identity*. We must have a clear understanding of the content of a particular national identity in order to predict whether it will lead to combat motivation in a particular conflict.

What Is Nationalism?

Following Anthony D. Smith, nationalism is "an ideological movement to attain and maintain autonomy, unity and identity on behalf of a population, some of whose members believe it to constitute an actual or potential 'nation.'"[50] However, in times of war, it is not simply the desire to attain and maintain political independence that motivates soldiers. It is the content of nationalism—the identity of the nation itself—that is a driving force. Nationalism as a concept offers only a broad framework for understanding a category of beliefs and identities that vary greatly across space and time.

Smith defines the nation as "a named and self-defining human community whose members cultivate shared memories, symbols, myths, tradition and values, inhabit and are attached to historic territories or 'homelands,' create and disseminate a distinctive public culture, and observe shared customs and standardized laws."[51] This definition is compatible with other scholars' definitions, such as Benedict Anderson's formulation of the nation as an "imagined community."[52] Nations can and have developed all over the world, but each nation is made up of a unique set of memories, symbols, and myths that form the identity of its members. Smith, following Hutchinson, refers to this as the "myth-symbol complex."[53]

I thus define *national identity* as *the shared memories, symbols, myths, and traditions of a nation*. Each national identity is made up of a unique narrative or set of myths and symbols that grants significance, meaning, and purpose to members of that nation.[54] Nationalism is an ideology that claims that nations—particularly *my* nation—should be politically autonomous.[55] National identity tells me who the members of my nation are, what I stand for as a member of that nation, and what we as a community should do with our political autonomy. Whereas an ideology is a commitment to a particular belief system, an identity is a set of expectations about who a person is, what they do, and how they relate to those around them.[56] National identity is one of many identities an individual may have, but it is particularly relevant to issues of international politics.[57]

For nationalism to influence political behavior, it must be filled out with the content of particular national identities. In fact, the general push of nationalism to prioritize and defend the nation tells us very little about the specific ways in which that will happen.[58] It is only through an examination of a particular national identity—the myths and symbols that give shape to this identity—that we can attempt to understand what effects nationalism may have on a state's military forces in a specific conflict. It is thus national identity rather than a generalized commitment to defending the nation (nationalism) that influences will to fight.

ARGUMENT

I argue that soldiers' will to fight comes from the commitment to the cause of the war and a sense of fighting for one's group. That commitment to the cause arises when political and military leaders are able to convincingly draw on the myths and symbols of national identity to frame the cause or the goals of the war. This is a two-step process: First, leaders must be able to draw on myths and symbols that resonate with their soldiers. This may not always be possible. In some states, for instance, there is no shared national identity, or soldiers' identities are different from those of their leaders. As Jason Lyall points out, ethnic identity can also be used as a marker to permit the state to repress certain groups, reducing their morale in combat.[59] In many other situations, however, there is a shared national identity among soldiers and between soldiers and leaders, and the myths and symbols of that identity are available as a rhetorical device on which leaders may draw.

Second, the soldiers must view the goals of the war as compatible with the national identity. This also may not always occur. In some cases, there may be a clear disjuncture between the goals of the war and the national identity. For example, while almost all nationalist leaders in India prior to WWII were opposed to fascism, they were also opposed to continued British rule in India. Fighting to preserve the British Empire, even from fascist threats in Germany and Japan, was therefore not compatible with Indian national identity. In other cases, elites may be able to frame the goals of the war as compatible with national identity initially, but events over the course of the war may cause the public (and soldiers) to reject that framing. The Vietnam War illustrates this dynamic. The American public and its soldiers initially supported the intervention as a war to protect democracy from communist expansion. But as the war dragged on and it became clear that South Vietnam was not a democracy and that American troops were committing war crimes, support for the war among the public and soldiers declined.[60] That said, it is nonetheless possible that elites will successfully frame the goals of the war as compatible with the national identity, and that this framing will last for the entire conflict. As we will see, such was the case for British soldiers fighting Germany during WWII.

Tying these elements together, if leaders can successfully frame the goals of the war in terms of nationally poignant myths and symbols, it will both spark group emotion and legitimize the war. On the battlefield, that emotion and sense of legitimacy will fuel sacrifice and risk-taking; in other words, motivate soldiers to fight as hard as they can. I explain this theory in detail in chapter 1.

PLAN FOR THE BOOK

I test the plausibility of my theory, as well as the alternative (yet complementary) arguments, through nine in-depth case studies of British imperial forces during the Second World War. I compare the varying will to fight among British, Indian, and Australian units fighting together in three major campaigns: North Africa 1940–42, Malaya 1941–42, and Europe 1941–44. These cases allow me to compare the different national units to one another in the same campaigns, and to themselves in different campaigns, capturing enlightening cross-case and within-case variation in will to fight.

The remainder of the book proceeds as follows. In chapter 1, I develop my theory in more detail and lay out my plan for observing the causal variables for each of the arguments I consider: national identity theory, small-unit cohesion, social structure, and threat. I also elaborate on how to observe the outcome of interest—will to fight—in order to measure it carefully in my cases. Chapter 2 explains the myths and symbols that make up the national identity of each of the groups under examination (British, Indian, and Australian) before the war and establishes the regime type and level of small-unit cohesion for each case. To be clear, this chapter describes the dominant myths and symbols of these identities but does not endorse or defend them as correct or appropriate. Neither is national identity synonymous with race or ethnicity. Chapter 3 focuses on the will to fight of each of the national units as they fight in North Africa, Chapter 4 on the fight in Malaya, and Chapter 5 details the campaigns in Greece and Italy in Europe. The conclusion considers the implications for international security scholarship and policy.

NOTES

1. Reese, *Why Stalin's Soldiers Fought*, 34.
2. Lepre, *Fragging*, 22.
3. See, for example, Brathwaite and Konaev, "War in the City;" Brooks, "Introduction: The Impact of Culture, Society, Institutions, and International Forces on Military Effectiveness"; Castillo, *Endurance and War*; Chacho, "The Influence of Ideology"; Lyall, *Divided Armies*; Reese, *Why Stalin's Soldiers Fought*.
4. Throughout the book I use the term *soldiers* as a catchall term for all combatants. This is in part for simplicity, but also because my cases focus on land combat. Aerial and naval combat differ in important ways from land combat and may require different explanations for will.
5. See, for example, Brathwaite, "Boys Will Be Boys?"
6. Kier, "Homosexuals in the U.S. Military"; Mackenzie, *Beyond the Band of Brothers*.
7. Brathwaite, "Effective in Battle"; Connable et al., *Will to Fight*, 27; Sarkesian, "Combat Effectiveness," 8.
8. Connable, 2.
9. Kellett, *Combat Motivation*, 6.
10. Brathwaite, "Effective in Battle," 4; Connable.
11. Simunovic, "The Russian Military in Chechnya," 66.
12. Gal and Manning, "Morale and Its Components," 370; Marshall, *Men Against Fire*, 158; Baynes, *Morale*, 92; Kellett, 7.
13. Biddle, *Military Power*, 48.
14. Lyall, "Forced to Fight," 88.
15. Castillo, 22; Reiter and Stam, *Democracies at War*, 65; Biddle, *Military Power*, 49. Grauer and Quackenbush, "Initiative and Military Effectiveness."
16. Van Creveld, *Command in War*, 270.
17. Ambrose, *Citizen Soldiers*, 66.

18. See, for example, Pawinksi and Chami, "Why They Fight?"; King, "On Combat Effectiveness in the Infantry Platoon"; Castillo,.
19. Connable, xiii.
20. Shils and Janowitz, "Cohesion and Disintegration," 306; Hauser, "The Will to Fight," 197.
21. Moskos, *The American Enlisted Man*, 156.
22. Baynes, 102; Shils and Janowitz, 284; Wong et al., *Why They Fight*, 9–10.
23. Hauser.
24. Kier, "Homosexuals in the Military," 17; Fuller, *Troop Morale and Popular Culture*, 44.
25. Lake, "Powerful Pacifists," 30; Biddle and Long, "Democracy and Military Effectiveness," 531.
26. Biddle and Long, 531.
27. Ambrose, 37; Reiter and Stam, 4.
28. Reiter and Stam, 4.
29. See, for example, Ferguson, *The Pity of War*; Reese, "Surrender and Capture."
30. Grauer, "Why Do Soldiers Give Up?"
31. Reiter and Stam, 66.
32. Morrow, "The Institutional Features of Prisoners of War Treaties."
33. Grauer, 630; Lehmann and Zhukov, "Until the Bitter End?"
34. See, for example, Mearsheimer, *Why Leaders Lie*, 72; Van Evera, "Hypotheses on Nationalism and War," 10; Desch, *Power and Military Effectiveness*, 60–61.
35. Bartov, "The Conduct of War," S36.
36. King, *The Combat Soldier*, 32.
37. Reiter and Stam, 79.
38. Baum and Potter, "The Relationship Between Mass Media, Public Opinion, and Foreign Policy"; Walsh, "Precision Weapons, Civilian Casualties, and Support for the Use of Force"; and Johnson and Tierney, *Failing to Win*.
39. Moskos, *Soldiers and Sociology*, 3–4.
40. Henderson, *Cohesion*, 4, 56–57.
41. Moyer, "Ideology and Military Systems," 108.
42. King, *The Combat Soldier*, 73.
43. Castillo, 28.
44. Bartov, *The Eastern Front*; Fritz, *Frontsoldaten*. See also Kershaw, "How Effective Was Nazi Propaganda?"
45. See, for example, Cederman et al., "Testing Clausewitz;" Posen, "Nationalism, the Mass Army, and Military Power."
46. Castillo, 11.
47. Kocher, Lawrence, and Monteiro, "Nationalism, Collaboration, and Resistance"; Hutchinson, *Nationalism and War*.
48. Malesevic, "The Structural Origins of Social Cohesion," 741.
49. Lynn, *The Bayonets of the Republic*, 21; Reese, *Why Stalin's Soldiers Fought*, 106.
50. Smith, *Ethno-Symbolism*, 61.
51. Smith, 29.
52. Anderson, *Imagined Communities*, 6.
53. Smith, 24.
54. Kaufman, *Modern Hatreds*, 25; Bottici and Kuhner, "Between Psychoanalysis and Political Philosophy," 98.

55. Smith, 61.
56. Connable, 52–54.
57. Bloom, *Personal Identity*, 50.
58. Smith, 99.
59. Lyall, *Divided Armies*, 5.
60. Lepre, 11; Hauser, 189.

CHAPTER 1

Explaining Will to Fight

Why are some militaries more motivated in battle than others? The answer to this question is essential for understanding war outcomes, conflict processes, and even military power in general. In this chapter I lay out national identity theory as an explanation for will to fight. National identity theory argues that when a military holds to a common national identity, and when the goals of a war are perceived by soldiers in that military to match their national identity, the military will have high will to fight. Soldiers who identify the goals of their battle with emotionally resonant symbols of national myth will have strong morale, take initiative, and be disciplined in battle. In contrast, when a military lacks a unified national identity, or if soldiers struggle to see the legitimacy of the war they've been sent to fight or how it fits with their shared historical and cultural milieu, they will exhibit lower morale, poor discipline, and be less motivated to fight as a whole.

National identity theory does not compete with arguments that center on small-unit cohesion, democratic effectiveness, or threat to the nation to account for the differences in levels of combat motivation we observe throughout time and space. Rather, it builds on this previous work and adds elements of scholarship on the role of narrative and emotions in political behavior to offer a unified and dynamic view of national identity—*the shared memories, symbols, myths, and traditions of a nation*—as the process by which nationalism influences will to fight.

NATIONAL IDENTITY AND WILL TO FIGHT

Psychology tells us that human beings need identity. We need to know who we are in relation to others and what our role is in our communities, whether family, workplace, or nation.[1] Individuals possess multiple identities that are constructed and reconstituted through interaction with the social and political environment in which they find themselves—mother or father, coal miner or software engineer, Catholic or Protestant, and so on. Even cities are often described in terms of the identities of their residents. Pittsburgh, for example, is generally seen as a gritty blue-collar, working-class "rustbelt city"—an identity that has held up despite the growth of its dynamic and lucrative biomedical sector. Additionally, national identity may or may not correlate with ethnic or racial identities. American national identity has at times made room for some variety of ethnic identities while excluding others. National identity may include racist myths (as in some of the cases in this book), but that is not necessary.

National identity is one such identity, constructed out of the myths and symbols, memories, and traditions that permeate our social and political environment. Anthony Smith calls this the "myth-symbol complex" of ethnic or national identity.[2] Myths tell of past glories, decry past humiliations, and proclaim future purposes.[3] Symbols are the objects, acts, words, or rituals that bring to mind and enact those myths.[4] Singing "The Star-Spangled Banner" at American football games, for example, is a symbolic and participatory act that recalls to mind the American national myth of defending freedom from supposedly despotic attacks (the War of 1812 and others). And when combined with symbolic demonstrations of American military might such as military flyovers, it evokes strong emotional responses of pride and confidence from people who otherwise have little reason to ponder their national identity right before kickoff. Schools, religious and civil society organizations, professional associations, the media, and political leaders, then, all help to construct and perpetuate a national identity by retelling myths and evoking symbols of that particular identity. Myths and symbols of national identity help people to feel part of a group but can also be used to exclude people from that group. As Jason Lyall points out, identity defines who you are as well as what can be done to you by political leadership.[5]

During times of war, national identity can become a source of emotional motivation for soldiers in battle. Where there is an existing, shared national identity, leaders can draw on those myths and symbols to frame the goals of the war and engage soldiers' emotional connection to the nation. However, soldiers (and the general population) must view the goals of the war

Table 1.1: Predicted Interactions Between Identity and War Goals

	Strong Fit with Goals	Weak Fit with Goals
Unified National Identity	High will	Adequate will
Weak/Contested National Identity	Adequate will	Poor will

as compatible with those myths and symbols. Leaders cannot make up new myths and symbols nor twist those that do exist beyond recognition. Jasen Castillo rightly points out that nationalist rhetoric does not always work to motivate soldiers.[6] It is only when the myths and symbols are widely recognized and the goals of the war appear to be compatible with those myths and symbols that national identity serves to motivate soldiers. Omer Bartov argues that Nazi propaganda on the Eastern Front was effective in motivating soldiers in part because it put forward a goal that aligned with German identity and mythology to which soldiers had been exposed all their lives. In fact, antisemitism and anti-Slav racism in Germany predated the First World War.[7]

Myths and symbols, when they resonate with the intended audience, spark group emotion such as pride, fear, anger, and empathy. The war is legitimized if leaders are able to successfully frame the goals of the war in terms of those myths and symbols. That emotion and legitimacy in turn leads to sacrifice and risk-taking on the part of soldiers in battle; in other words, it leads to will to fight. Table 1.1 shows the different possible outcomes of this process.

Drawing on a wide range of scholarship, from anthropology to psychology to literary theory, the following sections show the logic of each step in my argument—connecting emotions, national identity, and behavior on the battlefield.

Emotions and the Study of International Politics and Conflict

Challenging the rationalist basis for much of international relations theory, feminist and constructivist scholars point to the role of emotion in political behavior in general, and in violence and conflict more specifically. Neta Crawford highlights the role of emotions in theories of international politics, emphasizing in particular the importance of fear and empathy in explaining states' international behavior. Fear can become institutionalized in practices that lead to conflict, while empathy can increase the possibilities for cooperation.[8] Roland Bleiker and Emma Hutchinson encourage scholars to expand their methodological approaches to account for emotion in politics, emphasizing the importance of art in understanding political perceptions and dynamics.[9] Jonathan Mercer and Sara Ahmed separately argue that emotion is socially constructed and influences group behavior in ways inseparable from rational decision-making.[10] As Crawford writes, "once fear is aroused, there is no simple way to disentangle thinking from fear and fear from thinking. So too with empathy."[11] Working explicitly on the role of emotions in ethnic violence, Roger Petersen argues that emotions influence individual calculations of preference and willingness to take risks.[12] Drawing our attention to the conduct of war, Samuel Zilincik argues that emotions are key to understanding military strategy.[13]

These scholars draw attention to the limits of rationalist explanations of international politics and to the importance of both individual and group emotions in political decisions. I build on this scholarship to better understand soldiers' decisions and behavior in war, and particularly the role of emotions in shaping the will to fight through boosting (or sapping) morale, fortifying (or fracturing) discipline, and inspiring (or dampening) initiative on the battlefield. References to myths and symbols call up emotions attached to particular identities without necessarily requiring individuals to reflect carefully on those identities. What connects emotions to political and battlefield behavior are the national myths and symbols evoked by leaders as they relay the goals of the war to the public in general and the military more specifically. This is the topic to which we turn next.

Myths, Symbols, and National Identity

Myths abound in every area of social life. They help us to make sense of the world around us; they provide significance and meaning to our lives, telling us where we have been and where we are going.[14] Myths inform religious communities, ethnic groups, and nations. This book is concerned with political myth, which Chiara Bottici and Angela Kuhner define as "the work on a common narrative, which grants significance to the political conditions and experiences of a social group."[15] As distinct from a sacred myth, which tells the story of the relationship between god and god, or god and humanity, political myths focus on relationships between humans.[16]

A myth is more than just a story. Rather, it is a narrative that grants significance and meaning to a particular group and is used to help understand and act on the political conditions in which the group lives.[17] Myths explain who the group is, where it came from, and confer on it a sense of purpose.[18] Political myths in particular explain the origins and purpose of the political community or nation.

Myth is not necessarily either empirically true or false. Rather, it is a socially constructed narrative that claims to be historically true. It does not spring completed from nature, but neither is it purely the product of elite manipulation. Rather, myth is constructed by the interaction between elites and audience.[19] National identities and the myths that define and construct them are neither empirically true primordial realities nor false elite narratives. Instead, national identities are continually constructed through the interaction of the people and the elite. For example, public school history textbooks in the United States have long been a site of contestation regarding the myths of national identity. After the US Civil War, the United Daughters of the Confederacy worked hard to enshrine their narrative regarding the causes of the war in history textbooks throughout the United States. That narrative downplayed the role of slavery in the South and focused instead on honor and Northern aggression. African American communities have challenged that narrative and the "lost cause" myth that it helps to perpetuate.[20] Myths and symbols surrounding the Civil War have long shaped American understandings of who is American and who is not, and have changed over time.

The myths and symbols that make up national identity are formed socially as communities seek to answer questions of purpose and meaning in the world and to respond to shifts in political context.[21] Moreover, the process is cumulative and often subconscious. As Bottici and Kuhner note, "The work on a political myth is a process that can take place in the most diverse settings: speeches, arts, both visual and non-visual, rituals, social practices, and so on."[22] Individual members of society absorb the political myth as part of living within the community and participating in society but may not be able to explicitly articulate the myth. This is what distinguishes national identity from an ideology; national identity is communal and in part subconscious, whereas ideology is an explicitly articulated political system.

Myths are specific to each community.[23] Identifying national myths requires looking for the narratives that people tell about themselves, the stories they use to explain particular events or to motivate specific policies. Scholars have suggested three characteristics that distinguish a myth from a simple narrative: (1) it reproduces significance, (2) it is shared by members of the group, and (3) it addresses the specific political conditions in which the group lives.[24] An example of a narrative that meets these criteria for myth is American exceptionalism, which tells a story about the origins of the American nation and its significance in the world. This myth is shared widely by the members of the American nation and is used to understand issues as diverse as continental expansion and the 9/11 terrorist attacks.[25] By applying these three criteria, we can examine the specific myths that make up different national identities and ask how those myths might shape behavior.

Myths are often called up or referenced by shorthand, through the use of symbols. Symbols are rituals, ceremonies, visual images, locations, or music that call to mind the national myth.[26] Symbols have meaning only because of the political myth that they reference. For example, Bottici and Kuhner argue that an illustration of a veiled Marianne in a French newspaper cartoon has meaning because (for the French) Marianne is a recognized symbol of the French myth of national secularism and liberty that defines who is French and what it means to be part of the French nation. The veil is then a symbol for religious oppression, defining those outside the French nation and threats to that nation. They note, "one must not only be aware that the woman represented is called Marianne and that she is the symbol of the French Republic, but also have been exposed, more or less subconsciously, to the work of elaboration of these two symbols (Marianne and the veil) that took place in the specific French context to capture the full significance of the icon."[27]

The same symbol would not have gotten its message across had it been published in a newspaper in Indonesia or Brazil, and neither does it explicitly reference identity. Symbols have significance because of their connection to political myths that define an identity and because the viewer understands and affirms that connection, though as Bottici and Kuhner point out, that affirmation may not be consciously articulated. In the US context, this process is sometimes referred to as "dog whistle" politics, generally when politicians use symbolic language to call up racial fears and prejudice without actually referring to race. Such language would not have the same effect outside the United States. Elite use of particular symbols in political rhetoric must therefore be understood specifically within the context of the dominant political myths that define the nation and its members.

Myths and the symbols that bring them to mind give content and meaning to a particular national identity and evoke an emotional response from their audience. In 2016–17 the backlash against NFL players who knelt during the national anthem as a form of protest against police brutality suggests how emotionally important and politically salient such symbolic practices can be. People (including Vice President Pence) walked out of games in (counter)protest. Some even burned jerseys and threatened boycotts to demonstrate their disgust with an action they perceived as denigrating their identity as Americans.[28] National identity is thus constructed by myths and symbols that explain who is part of the nation, who is not, and what it means to be part of the nation. In this example, Black Americans taking a knee in silent protest during the national anthem was viewed as an attack on white American identity, though few of those white Americans explained their emotional reaction in terms of the white supremacist roots of the national anthem. One fan said, in a post of a video showing a burning Colin Kaepernick jersey, "Welp [sic] when you don't rise for the flag to show your respect for the men and women who fought and continue to fight for our freedom you start losing fans."[29] Though Kaepernick's protest was against police brutality, it was read through the lens of the symbolic association between the national anthem, the flag, and the US military and resulted in political actions. These ideas help pave the way for my argument that by drawing on myths and symbols, political leaders can mobilize members of the nation for action, evoking an emotional response that leads soldiers to take risks and make sacrifices on behalf of the nation, even if they do not explicitly reference their identity.[30]

Myths, Symbols, and War Goals

International relations scholars recognize that the myths and symbols that make up national identity are important for understanding international politics, foreign policy decision-making, and conflict behavior and processes. Ronald Krebs, for instance, examines leaders' use of stories to explain important world events and promote specific policies, finding that policies at odds with dominant narratives draw few supporters, while legitimacy is conferred on those policies in line with the dominant national narrative.[31] William Walldorf also argues that stories and narratives shape policy choices, constraining leaders' policy options.[32] Stuart Kaufman argues that ethnic conflict is caused in part by leaders who call on existing, emotionally powerful myth-symbol complexes that paint different ethnic groups as enemies or threats.[33] Relatedly, Jeremy Ginges and Scott Atran find that narratives centered on religious identity are important explanations for extremist violence, justifying participation in violence even when the expected utility is low.[34]

According to Kaufman, leaders can draw on these myth-symbol complexes to mobilize their people for conflict, but they cannot invent them—once again highlighting the importance of grounding these political appeals in national narratives that remain meaningful for the audience at hand.[35] Indeed, a key element of national identity theory is that elites cannot simply "make up" a convenient national myth or symbol in order to legitimize a particular policy. Myths and symbols must resonate with the intended audience to be effective. Scholars of securitization note that identity can limit the effectiveness of elites' securitizing rhetoric.[36] While myths are constructed, they are socially constructed, and change must therefore involve society, or some elements of it. An example is recent debates about how to teach history in the United States and United Kingdom. Historians and other academics, along with political activists, have put forward retellings of national histories that include attention to the settler-colonial and imperialist practices of both states. In the United States, efforts to include greater attention to the role of slavery and racial oppression in the development of the country has resulted in a significant backlash, with counter-histories put forward and laws passed banning teaching "divisive" concepts.[37] Efforts to shift the myths and symbols of US national identity have been embraced by some members of that nation but adamantly rejected by others.[38]

Equally important, policy and myth can clash, resulting in rejection of the policy and the politicians promoting it. Even long-standing myths that are generally accepted cannot always legitimize particular policies; in fact, they can be used to mobilize against those policies. Just as myths are constructed through the interaction of the people and the elite, framing particular issues in terms of a particular myth requires acceptance from the audience.[39] It is not enough for a leader to simply reference the myth; the audience must find the link between myth and policy plausible and convincing.

Kaufman points out that myths and symbols associated with ethnic or national identity help people to interpret events, but events themselves cannot rewrite myths and symbols.[40] Similarly, Walldorf notes that certain events and policies are more compatible with a particular narrative than others. During the 1980s, President Reagan advocated for "constructive engagement" with the apartheid South African regime, drawing on a set of American myths about the role of the free market in democratization and painting the African National Congress as a violent terrorist organization endangering the Black population. But the violent behavior of the apartheid regime belied this narrative, and American public support for divestment and sanctions grew.[41] Additionally, leaders can and do lie about the goals of a war. The Iraq War initially had significant popular support in the United States in part because the Bush administration claimed it was necessary to end the threat of weapons of mass destruction the country was developing. When it emerged that the claimed weapons program did not exist, support for the war decreased.[42] Lying may work for a time, but it is not sustainable in the long run.

In sum, elites can and do work to shape national identity, introducing new myths and symbols or reinterpreting those already part of the national lexicon.[43] However, these efforts take time and include audience response and interaction. The point of mobilizing myths and symbols to motivate an army is to draw out the emotional response to those symbols and channel it toward combat. That emotional response is tied to the absorption of an identity and recognition of the associated symbols. New or unfamiliar myths or symbols are unlikely to successfully trigger emotion. Elites may successfully tie events to national myths and symbols, but existing myths and symbols must make that plausible. For example, Jarrod Hayes argues that Tony Blair successfully drew on Britain's democratic identity to convince the public that Saddam Hussein was a threat to global peace but failed to convince them that the appropriate response was a war unsanctioned by the United Nations. Britain's commitment to UN authority as part of its democratic identity meant that Blair failed to mobilize public support for the war, and he paid a political cost.[44]

Because myths and symbols of national identity are not infinitely malleable, leaders are limited in what kind of policy choices can be justified by appealing to these shared historical and cultural experiences and memories. This is particularly true when it comes to defining and conveying the goals of a war to the public and the military drawn from that public. Leaders may attempt to frame a conflict in terms of national identity and fail. As we will see in the chapters that follow, this was the case for the British in trying to frame the war for Indian soldiers. Indian myths and symbols did not lend themselves to justifying defense of the British Empire, and as such, failed to consistently inspire high levels of combat motivation. In other words, the specific goals of the war limit the options that leaders have for promoting the war through nationalism and can therefore be viewed as a key element of the link between national identity and will to fight.

Similarly, events over the course of the war may change leaders' ability to frame the conflict in terms of national identity. While Kennedy and then Johnson were at first able to frame the war in Vietnam as a fight to defend freedom and democracy from a totalitarian menace (evoking important elements of US national identity), events over the course of the war called that framing into question. Eventually, it became impossible to square American myths and symbols with the violence against civilians, such as the My Lai massacre, that contradicted those myths and symbols on the ground.[45]

That is not to say that a policy's success or failure determines compatibility with national identity. In fact, Jeremy Ginges and Scott Atran argue that when group members are committed to the cause out of belief that it fits with group myths, perceived efficacy of a particular policy or action is less important than the fact that it fits with those myths. They write, "perceived righteousness but not perceived effectiveness predict willingness to take part in political violence."[46] Individuals support policies and even participate in violence when they believe that such actions are legitimate, even if they are unlikely to succeed. And to a great extent, legitimacy comes from the perception that an action or policy is compatible with the myths and symbols that make up national identity.

Myths, Symbols, and Will to Fight

National myths and symbols are not simply rhetorical devices. To have an impact on will to fight, they must have an impact on behavior. But do soldiers actually care about the larger politics of their battles? Governments certainly seem to think so, with many spending enormous resources to convey the causes of the war not only on the home front but also on the front lines. Both Axis and Allied powers in WWII developed a range of media propaganda aimed at encouraging their own soldiers to fight harder and to convince opposing soldiers the fight was worthless. Hollywood made numerous films for the British and American governments aimed at the general public as well as soldiers explaining why they were fighting; Frank Capra's *Why We Fight* series was among the least subtle examples.

Still, a common claim when discussing motivation in battle is that soldiers are generally apolitical. They want to do their job and get home. Indeed, studies of WWII soldiers—particularly Allied soldiers—find that they expressed little interest in the politics of the war. The most common motivation given by US soldiers in surveys was the desire to get home.[47] David French argues that British soldiers in that war showed little interest in politics and rarely exhibited real bitterness or hatred toward the Germans.[48]

While few would disagree that most soldiers want to get home as soon as possible, the idea that combatants are generally uninterested in the goals of the war overstates the case. The period of reconstruction after the US Civil War may explain part of the reason US soldiers were so uncomfortable with wartime patriotic expressions. Thomas Bruscino argues that efforts to put the politics of the Civil War and Reconstruction behind it led the country to discuss the war in terms of common sacrifice rather than cause. This tendency was reinforced by disillusion with patriotism after World War I.[49] Similarly, French notes that British soldiers were deeply suspicious of patriotic appeals and even the government's truthfulness after the propaganda of WWI.[50]

Although WWII-era US soldiers were unlikely to describe their fight in terms of high political goals, there is still evidence that they were concerned with the goals of that war. Tania Chacho, in digging deeper into surveys and studies of US paratroopers in that war, finds that soldiers who expressed an identification with the aims of the war were more motivated to go into combat.[51] Omer Bartov finds significant evidence for soldiers' desire for news and explanation of war aims in his study of Wehrmacht units on the Eastern Front.[52] These studies provide support for my argument that soldiers' identification with and understanding of the goals of the war may contribute to their will to fight. Nonetheless, whether soldiers consider politics is a real question, which must be addressed in my cases.

The emotional response to invocation of national myths and symbols may not involve a conscious reflection on identity. To return to the example of NFL fans upset by Colin Kaepernick's protest during the national anthem, those who burned jerseys or walked out of games did not always invoke their American identity. They referred to failure to respect the military, Kaepernick failing to love America, or to the fact that he was being paid $126 million to play football.[53] Like these fans, soldiers may not articulate their response to invocations of national myths and symbols in terms of their national identity, but the emotional response may be there nonetheless. There will be no emotional response at all, though, if they do not pay any attention to the politics or goals of the war.

For the purposes of the theory's development, assuming that soldiers do pay attention to the politics of the war in some capacity or another, how do we get from elites invoking myths and symbols of national identity to justify war, to soldiers' willingness to engage in battle? There are three ways that the compatibility between national myths and symbols and the goals of the war translates into soldiers' willingness to fight: increasing a sense of communal belonging, shaping perceptions of legitimacy, and generating intentions to act.

The myths and symbols of national identity are emotionally laden and calling on them when discussing a particular policy issue evokes feelings such as pride, fear, anger, and solidarity among the audience.[54] Using symbols that arouse emotion based on group identity can spark an emotional response even in those who are not directly affected by a particular policy by suggesting that their identity is affected. That emotion can in turn lead to stronger group identification, creating a cycle of identity and emotional commitment.[55] Shared emotional response to a situation can strengthen group affiliation.[56]

Emotions also shape the way groups perceive the world around them, legitimizing particular sets of policies and actions. Wendy Pearlman argues that "emotions easily bump mundane facts out of awareness," focusing attention on the facts related to the emotional response at the expense of nonemotional events.[57] Kaufman notes that emotions contribute to intuition; an emotional response to information means that information is more likely to produce strong memories and the type of emotional response helps to evaluate information: "If something feels good, then it is good."[58] In addition to shaping perception, emotions can legitimize actions. The emotions sparked by understanding the September 11, 2001, terrorist attacks through a set of myths and symbols constructing it as a conflict between good and evil legitimatized the invasion of two countries and violations of the Geneva Conventions.[59] Legitimizing acts of violence is particularly important for soldiers who are being asked to violate the basic social imperative against killing others.[60]

Finally, emotions influence behavior. Petersen notes that emotions can act as a "switch," selecting which among a set of basic desires one will pursue: "An individual may value safety, money, vengeance, and other goals, but emotion compels the individual to act on one of these desires above others."[61] People have different, often competing preferences, and emotions help to determine which will win out in a given situation.[62] Indirectly, the emotions evoked by appeal to identity provide motivation and sustaining energy for individuals to take risks. Not only do the emotions legitimize particular actions, they also provide the impetus for individual and group action.[63]

These three ways in which emotions linked to the myths and symbols of national identity can impact soldiers' will to fight are nonexclusive and in fact often interact. Atran's work, for instance, alludes to how the sense of communal belonging amplified by appeals to identity myths can generate imperatives for action. Atran argues that when groups rally around "sacred values," such as those found in a national myth, they are willing to accept high levels of sacrifice.[64] These values provide the emotional energy necessary to risk one's life fighting for the cause. He attributes the influence of emotion, evoked by myth, to an evolutionary imperative to preserve the group. Bloom argues for a more individual source of this influence, noting the psychological necessity of identity and the need to defend that identity when threatened.[65] Whatever the ultimate source, myth evokes emotion that makes group members willing to sacrifice themselves in support of the identity that myth defines.

The Warsaw Ghetto uprising during World War II provides an important example of the way these three effects of emotion can influence motivation to fight. When Jews took up arms against the Nazis in the spring of 1943, they did so knowing that they would fail. They were outgunned and outnumbered. Rachel Einwohner writes, "The main goal of the resistance, therefore, was not necessarily to beat the SS troops and secure safe passage out of the ghetto. Instead, it was to act honorably."[66] She argues that they saw resistance, even when doomed, as helping to restore honor to their identity and dignity to the Jewish people as a whole.[67] The emotions of pride and honor that arose out of Jewish identity increased a sense of solidarity within the group, prioritized violent resistance over potential alternatives (however limited), and increased group members' willingness to take risks.

Most of the above discussion focuses on the process by which invocation of myths and symbols can lead to an emotional response and thus action by individuals. But war is not fought by individuals as individuals. Battles are fought by groups of individuals—squads, battalions, divisions—who must work together to engage the enemy. How many of the individuals that make up those units must believe that the goals of the war match their national identity in order for the unit as a whole to be motivated? What level of disagreement is possible before the unit breaks down?

The literature on small-unit cohesion demonstrates that military units can and do work to develop a sense of collective purpose in order to achieve strategic goals. The concept of task cohesion is especially relevant here. A number of studies have demonstrated that while small-unit cohesion in the form of social cohesion may provide a source of combat motivation, without a commitment to the overall goal—task cohesion—it can also do the opposite.[68] Soldiers with strong emotional bonds who prioritize each other's survival may collectively decide not to fight rather than risk their comrades' lives.[69] On the other hand, when soldiers have a common national identity and believe the goals of the war align with that identity, military leaders can use that motivation to develop task cohesion. In that way national identity theory compliments small-unit cohesion theory by explaining one factor that contributes to the development of task cohesion and thus unit performance in battle.

Moreover, Tania Chacho, Charles Moskos, and others have demonstrated that soldiers' motivations are multiple. Survival, personal relationships, honor, and fear of punishment can and do coexist with commitment to cause or political ideology.[70] National identity theory provides a model for the process by which a soldier's sense of group membership translates into combat motivation through task cohesion but does not argue that it is the only source of motivation.

This still does not give us a clear number of soldiers who must identify with the group in order to create that task cohesion. Theoretically, I would expect that "most" soldiers in the unit would need to identify with the group though—as discussed above—they may not be able to articulate that identity explicitly. The more soldiers share a common identity, the less explicit it might in fact be as it becomes taken for granted. I expect that military units' identities are roughly representative of the identities of the communities from which their members come. In other words, if a battalion is made up of Sikh and Hindu soldiers, then that unit is likely to share the general sense of those communities as to whether the goals of the war align with their national identities. Similarly, if a battalion is made up of British soldiers, we should expect the unit to share the general attitude of the British public to the relationship between the war goals and the myths and symbols of British identity. There will inevitably be some soldiers who reject their national identity or disagree with the majority as to its compatibility with the war goals, but I expect that to be roughly proportionate to such attitudes in the population as a whole. The more disagreement there is among the general population, the more disagreement I expect to see within military units and the less motivation that unit is likely to show in battle. Chapter 2 is dedicated to explaining the identity or identities that prevail in each of the countries under examination, in order to establish the identity communities from which soldiers come. I try to capture the continuous nature of motivation in my measure of will to fight—high will, adequate will, and poor will.

Impact on Morale, Discipline, and Initiative

To reiterate (and as suggested in table 1.1), when leaders invoke national myths and symbols in order to justify a war and mobilize society to participate in combat, the society determines whether they believe the goals of the war to be compatible with those myths and symbols. If the goals are determined to be incompatible, national identity should not have a positive influence on combat. In essence, evocations of identity would not work in that case. But if society in general and the military by extension find the goals of the war to be compatible with their national identity, then it should have the emotional effects of increasing communal belonging, shaping perceptions of legitimacy and generating willingness to act described above. Parsing out how this perceived congruence between war goals and national identity shapes combat motivation requires us to consider more closely the constitutive parts of will to fight—morale, discipline, and initiative.

Having a shared national identity can increase morale by providing soldiers with a source of emotional connection to the community and a commitment to the cause. Additionally, the ways that nationalism shapes perception help soldiers to see their circumstances as part of a noble act of sacrifice, rather than as a miserable waste. While soldiers should not be expected to take pleasure in their difficulties, national identity and commitment to the cause shape their attitude toward their situation. When members of the Kurdish Women's Protection Units (YPJ) were fighting Islamic State of Iraq and Syria (ISIS) in Kobani, Syria, they faced a well-armed and experienced enemy with very few military resources of their own. Their leaders reminded them of their goals; defending their Kurdish community and avenging the Yazidi women ISIS was brutalizing. YPJ members not only fought for the right to fight ISIS, but they risked torture and death in what appeared to be a losing fight in Kobani. They were willing to make such sacrifices in part because of an emotional connection to other members of their nation.[71]

That sense of communal belonging and responsibility to the home community can also strengthen discipline. Calling on myths and symbols of national identity helps to evoke group emotion among soldiers, which ties the battle to defending the security of their identity. By creating an emotional commitment to the goals of the war, national identity promotes obedience without recourse to threats. Additionally, if myths and symbols shape soldiers' perceptions of the battle and the enemy, they are likely to see greater security in complying with orders to ensure victory rather than avoiding battle. The pride and anger that identity evokes also make the perceived efficacy of the action taken less important than the fact that it fits with the group's myths. Even when soldiers might see victory as being out of reach, the importance of fighting for the nation outweighs any tendency toward indiscipline. North Vietnamese prisoners of war reported being motivated to fight for reunification and removal of foreigners from their homeland despite also acknowledging they could not defeat the American forces materially. Rather, they believed they were more committed and thus would outlast the Americans.[72]

Soldiers whose national identity is compatible with the goals of the war are also more likely to take initiative to achieve those goals. The emotions sparked by national identity—courage, sacrifice, and belonging—elevate the significance of victory over personal security, encouraging soldiers to take risks. Because they are not reliant on training alone for motivation, they are willing to try new tactics when those they were trained to use fail.

Emotion sparks intention to act, leading soldiers to be willing to seek out engagement with the enemy and even make sacrifices to achieve victory and uphold their national identity. The story of Zvika Greengold, an Israeli tank officer during the 1973 Arab–Israeli War, helps illustrate these dynamics. Driving a lone tank on the Golan Heights as the Syrian Army invaded, he fought throughout the night, leading the Syrians to wrongly conclude they were facing several enemy tanks. He later cited a sense of responsibility to Israel, saying "I was not scared of dying. I was scared of failing."[73]

To sum up, soldiers who are motivated by the myths and symbols of their national identity to achieve the goals of the war will have high levels of morale even under dire circumstances, exhibit strict discipline, and demonstrate initiative by seeking out engagement with the enemy and trying new tactics. By drawing together scholarship on identity, myth, narrative, and emotion, national identity theory provides an explanation for the conditions under which, and the means by which, national identity motivates soldiers in battle.

METHODS

National identity theory argues that discourse in the form of myths and symbols of national identity influence human behavior. The project in this book is therefore one of causal explanation, though it recognizes the complicated—and not entirely observable—nature of the social world.[74] To account for that complicated reality, I adopt a pragmatic approach to methods—one of eclecticism, drawing on examinations of both discourse and observed behavior.[75] To establish national identity, I look to discourse of myths and symbols that permeated each society before the war (see chapter 2).[76] I also examine discourse from soldiers themselves in the form of letters home, troop newspapers, and unit war diaries. Determining whether that discourse influenced their behavior requires examining soldiers' actions in battle. Here I rely on military histories as well as war diaries and after-action reports.

Just as no single law governs human behavior, no single factor can fully explain variation in soldiers' will to fight. Will may be influenced by overall military capability, level of training, quality of equipment, and familiarity with the terrain. As such, I seek to hold as many important factors constant as possible using a controlled-comparison case selection approach. I thus look for variation in national identity among military organizations fighting under similar conditions. The British Imperial Army during WWII provides a useful set of such cases.

The British Empire fielded over 8 million soldiers during WWII, from the British Isles, India, South Africa, Kenya, New Zealand, and other parts of the empire and commonwealth.[77] Additionally, soldiers who escaped occupied Europe—Poles, French, Czech, and others—also fought with the British forces. Because the different national groups fought as distinct units (South African units as South Africans, Indian units as Indian, etc.), the British Imperial Army provides variation in national identity that allows me to examine the influence of different national identities and their relationship to war goals on soldier motivation. In keeping with the controlled-comparison approach, I selected specific national armies for examination that fought under the most similar circumstances. British, Indian, and Australian soldiers all fought as distinct national units yet followed British basic training standards and military doctrine, using the same technology and equipment. Moreover, in several instances, the three forces fought the same opponents in the same locations at the same time. By comparing units of different nationality who fought side by side in the same battles and campaigns, I can hold constant a variety of other factors (enemy equipment, topography, etc.) that might explain variation in will to fight and focus on the explanatory variables of interest.

Additionally, having two democracies and one non-democracy allows me to consider the influence of domestic politics on will to fight, while the recruitment and organizational policies of the British and Indian armies promoted small-unit cohesion and the Australian did not. Through careful within- and between-case analysis, I can thus evaluate the influence of these variables of interest on will to fight.

I select campaigns from each of the three major theaters of war where all three national groups fought. Chapter 3 considers the campaign in North Africa against the Germans and Italians. It begins in 1940 when the Germans arrive in Libya and ends in 1942 when the Australians leave and are replaced by the Americans. Chapter 4 covers the defense of Malaya against the Japanese from the initial invasion in December 1941 until the evacuation to Singapore Island in February 1942. I end with the evacuation because once the forces are consolidated in Singapore, it is not possible to adequately distinguish between different national units. Chapter 5 examines the defense of Greece (February–April 1941) and invasion of Italy (December 1943–January 1944), primarily against the Germans. The British, Australians, and Indians fought together in North Africa and Malaya but not in any battles in Europe (at least not in significant numbers), disrupting the controlled comparison in that theater. I thus compare Australian forces who fought against the Germans in Greece and British and Indian forces who fought the Germans in Italy.

Each case (North Africa, Malaya, Europe) is a major campaign (or campaigns, in the case of Europe) made up of a number of smaller battles. I chose to focus on campaigns rather than individual battles for several reasons. First, because I argue that national identity and war goals must be perceived as congruent to lead to motivation, I want to allow for the possibility of change over time. It is theoretically possible that soldiers may initially find the goals congruent with their identity but then discover that leaders lied about the war or that the goals change. In fact, the goals of the war do change in December 1941 when Japan attacks the eastern part of the empire. It is also possible that my theory is incorrect about the speed at which national identity changes; it is possible that it changes during the war. Considering campaigns rather than single battles allows for changes in identity to become evident if they occur.

Second, campaigns allow me to compare the national units to themselves over time. Did British soldiers perform well only at the beginning of the campaign but decrease in motivation by the end? Did Australian soldiers become more or less motivated over time? Campaigns allow for within-case comparison. Finally, and quite practically, each campaign is a collection of battles that I consider in detail. Conceivably, I could consider battles as the unit of analysis and each campaign as a set of battles.

Of the numerous nationalities that made up the British Imperial Army, I chose to focus on British, Australian, and Indian forces for several reasons. First, the British and Indian armies were the largest imperial contingents, and Australia was the fourth largest (following Canada). Second, these three forces were present together in significant campaigns, allowing for better comparison of the variation in their national identity. Finally, racism was a major confounding factor. Black soldiers from the African and Caribbean colonies were not allowed to command units and faced extremely different treatment than white soldiers or even Indian soldiers. While corporal punishment was banned in the British Army in 1881, African soldiers were publicly beaten for even petty crimes throughout WWII.[78] This legal variation in treatment makes it difficult to compare motivations across those groups. It does, however, mean that I must be careful in generalizing from these cases. Examining will to fight among British, Australian, and Indian soldiers allows me to consider the plausibility of my theory but not to establish with certainty its generalizability beyond these nationalities or this war.

These cases are rich in available data, making them excellent for testing national identity theory. High-quality historical scholarship has been conducted on each of these campaigns, and there is also rich sociological and anthropological scholarship on the national identity of each group, which allows for a deep exploration of the myths and symbols salient at the time of the war. I also had excellent access to primary source material both to double-check and to supplement the secondary literature on the subject. I make use of materials such as troop newspapers, censorship reports, unit diaries, after-action reports, and official histories to both explore soldiers' perceptions of the war goals and to assess their level of discipline, morale, and initiative in battle. The individual sources are discussed in more detail in the cases in chapters 3–5

The British Imperial Army was organized as follows: divisions (three brigades, about 16,000 men) > brigades (three battalions, about 3,500–4,000 men) > battalion (three companies, about 500–1,000 men) > company (about 100–200 men). While I consider the will to fight of units across this range (division-company), the focus of the cases is at the battalion level.[79] With respect to the three indicators used to assess soldiers' will to fight, the sources detailed above occasionally discuss discipline and initiative at the company level, but most aggregate to the battalion level. Importantly, morale assessments by officers and the Censors Office aggregate at the battalion or brigade level, making it difficult to determine morale at a company level. Additionally, while British and Australian divisions were made up of a single nationality, Indian Army divisions included battalions of British soldiers.[80] It is thus necessary to consider those battalions separately in evaluating the role of national identity in will to fight.

As noted in table 1.2, there is significant variation on the behavior of interest in these cases—level of will to fight. British and Indian forces vary in will to fight across the different theaters of the war, while Australian forces maintain a high level of will across all theaters. Importantly, while all national groups fought with high motivation in North Africa, there was variation between them in Malaya and Europe.

Table 1.2: Actual Levels of Will to Fight

	British	Indian	Australian
Malaya	Adequate	Poor	High
North Africa	High	High	High
Europe	High	Adequate	High

Measuring Concepts

Will to Fight

To measure these three key elements of will to fight in particular units, I ask the following questions:

Morale: Soldiers' attitude toward their situation. What are the levels of desertion in the unit? How do soldiers describe their situation? What are officer assessments of their soldiers' morale? As an inner state (attitude toward the situation or emotion regarding it), morale is difficult to observe. Morale is subjective and is a complicated concept to measure. Soldiers may write about it in letters home, but they may have reason to make light of their situation to their loved ones.[81] Officer evaluations are secondhand, but there is not a clear imperative for officers to misrepresent their soldiers' morale in one direction or the other. Officers may want their superiors to think the soldiers are happy with their leadership, but they may also seek to lay the blame for failure on their soldiers' attitude. Additionally, for six of the nine cases (all cases in Europe and North Africa), censorship reports were compiled by military censors. Based on what the censors determined were representative letters, reports assessing the morale of units in each theater of the war were compiled and sent back to government capitals as well as military headquarters in the relevant theaters. These censors had no personal interest in skewing the state of morale, although national and racial biases should be carefully considered in their reading.[82] None of these questions on their own fully reveals morale but taken in concert provide a reasonable sense of overall morale in military units.

Discipline: Soldiers following orders. Does the unit stay in the fight until ordered to retreat or disengage? Does the officer need to threaten or actually use violence to ensure the unit follows orders? If ordered to retreat, do soldiers do so as a unit and fall back to a designated location? Do soldiers use passive resistance to orders (deliberate slowdowns, carelessness with equipment, etc.)? The answers to these questions point to the level of internal discipline that soldiers have and how much external discipline must be imposed by the military in order to keep them engaged with the enemy. Soldiers who retreat without orders are less disciplined than those who retreat under orders from leadership. Units that need threats or use of violence to follow orders require greater resources than units where force is not necessary. Russian soldiers who sold their weapons to the Chechens during the Battle of Grozny, for example, may not have technically retreated against orders, but handing weapons to the enemy certainly indicates a lack of discipline and overall will.[83]

Initiative: Soldiers seeking out engagement with the enemy and trying new tactics. Do units organize counterattacks? Do soldiers volunteer for difficult assignments or take on jobs outside their own training? Do soldiers experiment with new tactics when those they were trained for fail? Do soldiers seek out engagement with the enemy? It is possible for a disciplined unit to follow orders to the letter but do so with little intention to actually defeat the enemy. Successfully defending a site they have been ordered to defend but then following up on an opportunity by counterattacking the retreating enemy demonstrates that soldiers are motivated by more than simple fear of punishment. Along similar lines, experimenting with new tactics when the old ones fail and finding new ways to engage the enemy demonstrate a strong commitment to the outcome of battle.

While there is no singular way to aggregate these three dimensions into a unified assessment of will to fight, it is nonetheless necessary to tie them together in order to compare between cases. Units that exhibit morale, discipline, and take initiative have high levels of will. Units that have morale and discipline but no initiative have an adequate level of will. A unit that is missing either morale or discipline has only poor will. Initiative is dependent on some basic levels of morale and discipline. Thus, a unit without morale or discipline would be unlikely to demonstrate initiative. Put somewhat differently, initiative is less essential to will in battle than high morale and good discipline.

National Identity

Turning now to measuring the causal factors of interest, recall the definition of national identity: *the shared memories, symbols, myths, and traditions of a nation.* Bottici and Kuhner define myth as a narrative that reproduces significance, is shared by group members, and addresses specific political conditions in which that group lives.[84] Symbols are shorthand images, rituals, or ceremonies that call to mind those myths.[85] I thus examine national narrative, important past events (or their retellings), traditions, and widespread beliefs about what sets the nation apart from others. Because narratives of war and sacrifice are important aspects of national myths and symbols, I examine each national identity prior to the onset of war to ensure national identity and war goals are not endogenous.

Because national identity is a collective phenomenon, I examine society as a whole and establish the national myths and symbols that inform group identity in that society as a whole. For the British and Australians this was generally a single identity for the country as a whole. For the Indians, there are multiple identities present across the country, requiring an examination of several sets of myths and symbols. I dedicate chapter 2 to explaining these national identities and their acceptance in society as a whole prior to the war. Because soldiers in each country were drawn from a wide section of society, I assume that most identify with the same myths and symbols as the rest of their society, even if this identification is mostly subconscious. I then refer back to those identities in the examination of the militaries in battle in chapters 3-5 to determine how elites drew (or failed to draw) on existing national myths and symbols to make the case for the war.

To establish the content of each national identity, I ask the following questions: What is the role of the homeland in the group's identity? How is it discussed or referenced in myths? What religious and political principles explain the group's conditions? How does the community believe it should govern itself? How does that people see itself relating to the rest of the world? What sets it apart? How do past wars inform national purpose and threats to the nation?

To determine how compatible the society believed the identity to be with the goals of the war, I ask the following questions: What did political leaders claim the goals of the war to be? How supportive of the war was society at large? Did soldiers discuss the goals of the war in letters home or among themselves? Did soldiers believe the war to be worthwhile? Polling was a new science at the time of the Second World War, but where available I draw on polling data to understand public attitudes toward opponents, war goals, and the war's progress.

Existing Arguments

Small-Unit Cohesion and Esprit de Corps

The commitment of soldiers to their unit, leaders, and military organization and the emotional bonds between soldiers are not easily observable phenomena. There is a great deal of debate regarding how primary-group cohesion is formed. Some argue that a basic level of homogeneity is necessary: some similarity in values, nationality, or language that allows for trust and bonding.[86] Others argue that group experiences can create a cohesive unit even without prior commonalities.[87] Soldiers who go through intensive training together have the same basic skills and trust one another.[88] Soldiers who go through combat together bond through shared difficulty, which reinforces trust and cohesion.[89]

Because I focus primarily on larger units than the small-unit cohesion argument envisions, I pay particular attention to secondary-group cohesion. Sometimes referred to as esprit de corps, the emphasis here is on creating a sense of loyalty to the unit through trust in one's comrades and leaders and pride in the unit itself.[90] It is not just cohesion within the platoon or company that motivates soldiers but a sense of those relationships fitting into a larger organization such as the regiment or battalion. Militaries can thus adopt organizational and training practices as well as traditions designed to build cohesion, in turn increasing soldiers' willingness to fight.

Studies of cohesion therefore use a number of indicators to measure the level of unit cohesion in a particular military. I look to the following indicators to determine the level of cohesion in the units under study: soldiers' own assessments of their unit's cohesion, unit training and replacement policies, officer training, length of time the unit had been together, rituals and history associated with the unit, and officer assessments of their unit's cohesion. Where possible, I track when units are reorganized during a campaign, if they break up and are combined with other units, and if their officers are replaced.

Small-unit cohesion arguments generally focus on the platoon or company level. My data is primarily at the battalion or brigade level, though at times unit diaries and reports are available at the company level. Wherever possible I use evidence from the smallest unit. I also track when battalions are combined. Shils and Janowitz, Siebold, King, and others who focus on unit cohesion generally agree that development of cohesion requires training together and a military rotation system designed to keep units together as much as possible so would expect combining units during battle would reduce cohesion. I focus especially on the organization and training of each military to determine if they are organized to promote cohesion.

If small-unit cohesion is the primary driver of will to fight, I would expect to see militaries organized around small-unit cohesion—units who train together, whose training is focused on combat drills, and who rotate into and out of combat as a unit—to show high will to fight. Units where that cohesion breaks down due to casualties or the chaos of battle (or was never formed in training) should show poor will to fight. Traditions and rituals associated with pride in the unit are also designed to increase cohesion and should be associated with will.[91] If this factor is not important, I should see some units with strong cohesion exhibit low will to fight and other units that have not been together long or lack a sense of esprit de corps demonstrating high will in battle.

Democratic Effectiveness

If democracies produce more motivated soldiers, democratic soldiers should be committed to winning the war and should also be confident in the support of the home front. I look to statements about domestic politics and leadership to assess these views. Moreover, democratic soldiers should be creative and self-motivated as their cultural individualism plays out on the battlefield. We should thus see more initiative from democratic than non-democratic soldiers. The literature does not distinguish between levels or types of democracies, simply that soldiers from democratic systems of all sorts are motivated in battle. I thus compare two militaries from democracies—Australia and Great Britain—with each other as well as with a non-democracy—India.

Threat

This theory argues that direct threat to the homeland and dehumanization of the enemy increase will to fight. In the cases I look for evidence that soldiers considered the level of threat to their homeland when at the front, why they thought so, and that those who believed the threat was high had more will than those that did not. I also look for evidence of dehumanization of the enemy. Reception of propaganda, letters home, and treatment of prisoners are all indicators of how soldiers viewed their enemy.

CONCLUSION

The argument developed in this chapter does not claim that cohesion, social structure, and threat do not influence soldiers' will to fight. Rather, national identity theory builds on these insights and develops a theory of how identity and political cause work together to encourage and strengthen cohesion in the face of certain threats. The next chapter lays out the myths and symbols of national identity in Britain, Australia, and India prior to the beginning of World War II. It also provides an overview of the organization of each army and its focus—or lack of focus—on the creation of small-unit cohesion.

NOTES

1. Bloom, *Personal Identity*, 33; Stern, "Why Do People Sacrifice for Their Nations?" 225; Castinera, "Imagined Nations," 43.
2. Smith, *Ethno-Symbolism*, 24.
3. Edelman, *The Symbolic Uses of Politics*, 6.
4. Edelman, 19. See also Castinera, 52.
5. Lyall, *Divided Armies*, 5.
6. Castillo, *Endurance and War*, 27.
7. Bartov, *The Eastern Front*, 88.
8. Crawford, "The Passion of World Politics;" Crawford, "Institutionalizing Passion in World Politics." See also Ariffin et al. (eds.), *Emotions in International Politics*; Hall, "We Will Not Swallow This Bitter Fruit."
9. Bleiker and Hutchinson, "Fear No More."
10. Ahmed, *The Cultural Politics of Emotion*; Mercer, "Feeling Like a State." See also Markwica, *Emotional Choices*.
11. Crawford "Institutionalizing Passion," 540.
12. Petersen, "Identity, Rationality, and Emotion."
13. Zilincik, "The Role of Emotions in Military Strategy."
14. Bottici and Kuhner, "Between Psychoanalysis and Political Philosophy," 98.
15. Bottici and Kuhner, 98.
16. Flood, *Political Myth*, 33.
17. Bottici and Kuhner, 99.
18. Flood, 34.
19. Esch, "Legitimizing the 'War on Terror,'" 364. For more on the importance of audience in constructing narratives, see Balzacq, "A Theory of Securitization."
20. McCarthy, "The 'Lessons' of the United Daughters of the Confederacy.".
21. Bottici and Kuhner, 99.
22. Bottici and Kuhner, 99.
23. Bottici and Kuhner, 105.
24. Esch, 362; Bottici and Kuhner, 99.
25. Esch.
26. Flood, 17; Bleiker and Hutchinson.
27. Bottici and Kuhner, 105.
28. Taylor, "'Burn the NFL."
29. Carissimo, "People Are Burning Colin Kaepernick Jerseys."

30. Bottici and Kuhner, 108.
31. Krebs, *Narrative and the Making of US National Security*.
32. Walldorf, *To Shape Our World for Good*.
33. Kaufman, *Modern Hatreds*.
34. Ginges and Atran, "War as a Moral Imperative," 2932.
35. Kaufman, *Modern Hatreds*, 30.
36. See, for example, Balzacq; Hayes, "Identity, Authority, and the British War in Iraq."
37. See, for example, LaCasse, "Divisive Concepts Ban Is New Hampshire Law."
38. See, for example, Hannah-Jones et al., "The 1619 Project"; President's Advisory Commission, *The 1776 Report*.
39. Kaufman, *Nationalist Passions*, 49; Krebs.
40. Kaufman, *Modern Hatreds*, 34.
41. Walldorf, *Just Politics*, 127; *The New York Times*, "Transcript of Talk by Reagan on South Africa" July 23, 1986
42. Rosentiel, "Public Attitudes Toward the War in Iraq: 2003–2008."
43. See, for example, Jachovic, "Reinforcing the National Identity"; Onuch and Hale, *The Zelensky Effect*.
44. Hayes.
45. Lepre, *Fragging*, 22.
46. Ginges and Atran, 2932.
47. Pfau, *Miss Yourlovin*.
48. French, "You Cannot Hate the Bastard Who Is Trying to Kill You."
49. Bruscino, "The Analogue of Work," 89 and 97.
50. French, "You Cannot Hate," 4.
51. Chacho, "Why Did They Fight?" 79.
52. Bartov, *The Eastern Front*, 77.
53. Boren, "Colin Kaepernick Protest Has 49ers Fans Burning Their Jerseys.".
54. Edelman, 5.
55. Mercer, 522; Ahmed, 10.
56. Van Ness and Summers-Effler, "Emotions in Social Movements," 417.
57. Pearlman, "Emotions and the Microfoundations of the Arab Uprisings," 391.
58. Kaufman, *Nationalist Passions*, 35
59. Esch, 365.
60. Cochran, "The Civil-Military Divide in Protracted Small War," 77.
61. Petersen, 398.
62. Pearlman, 391.
63. Petersen, 399; Kirke, "Violence and Political Myth," 286.
64. Atran et al., "Devoted Actors Sacrifice for Close Comrades and Sacred Cause," 17702.
65. Bloom, 52.
66. Einwohner, "Opportunity, Honor, and Action in the Warsaw Ghetto Uprising of 1943," 666.
67. Einwohner, 666.
68. See, for example, Kier, "Homosexuals in the U.S. Military."
69. See introduction for discussion of "fragging" in the US military.
70. See, for example, Moskos, *The American Enlisted Man*; Henderson, *Why the Vietcong Fought*; Chacho, "Why did they Fight?".
71. Tzemach Lemmon, *The Daughters of Kobani*, 89.

72. Donnel et al. *Viet Cong Motivation and Morale in 1964*, 33.
73. Lappin, "I Was Willing to Die to Stop the Syrian Advance."
74. Kurki, "Critical Realism and Causal Analysis in International Relations," 365; Archer, "Introduction: Realism in the Social Sciences," 199.
75. Kurki, 371; Archer, 200; Sil and Katzenstein, "Analytic Eclecticism."
76. As Margaret Archer notes, establishing the sequential nature of the interaction between structure (in this case national identity) and agent (soldiers) is essential. Archer, 202.
77. BBC, "World War II People's War."
78. Losh, "Britain's Abandoned Black Soldiers."
79. For a similar approach to levels of analysis, see Connable, *Will to Fight*, 25.
80. This practice developed after the Indian Army Mutiny to guard against rebellion among Indian soldiers. More on this practice in chapter 2.
81. Low levels of desertion may indicate commitment to the fight, but it might also mean that there is nowhere to which soldiers can flee. Desert and island warfare offer few escapes, whereas combat close to soldiers' home regions may offer a tempting alternative to the fight.
82. See chapter 5 in particular for such biases.
83. Simunovic, "The Russian Military in Chechnya," 81.
84. Bottici and Kuhner, 198.
85. Edelman, 6; Bleiker and Hutchinson.
86. Simmons, "Here's Why Women in Combat Units Is a Bad Idea" Shils and Janowitz, "Cohesion and Disintegration," 285.
87. Kier, 8.
88. Kinzer Stewart, *Mates and Muchachos*, 45.
89. Ben-Shalom et al, "Cohesion During Military Operations," 73; King and Bury, "A Profession of Love," 202.
90. Siebold, "The Essence of Military Group Cohesion," 287.
91. Fuller, *Troop Morale and Popular Culture in the British and Dominion Armies*, 44; Connable, 77.

CHAPTER 2

National Identity, Democracy, and Cohesion in the British Imperial Forces

National identity theory argues that the goals of the war must be compatible with the content of the national identity; this chapter explains the myths and symbols that make up each national identity prior to the onset of WWII. The case study chapters will refer back to those myths and symbols to establish whether soldiers believed that the goals they were fighting for matched their national identity. This chapter also establishes whether the military is organized around creating small-unit cohesion and whether the military comes from a democratic society. Threat is specific to each army and its enemy, and so is considered separately in each case.

I establish the myths and symbols of national identity through an examination of the historical, anthropological, and sociological scholarship of nationalism and national identity for each group. I focus on the period prior to the onset of World War II in order to account for the fact that war may influence identity. National identity theory argues that soldiers' identities will be generally congruent with the national identity of the society from which they are drawn, so this chapter establishes whether there was a generally agreed national identity or if there were competing identities, and asks the following questions for each identity: What is the role of homeland? How is it discussed or referenced in myths? What religious and political principles explain the group's conditions, and how does the community believe it should govern itself? How does the people see itself relating to the rest of the world? What is its purpose?

How do past wars inform national purpose and perception of threats to the nation?

To determine whether a military is likely to have high small-unit cohesion and esprit de corps, I examine their training, rotation, and deployment systems. I also ask whether the military encourages traditions and rituals to develop pride and loyalty to the unit. To establish the level of democracy, I use the Polity V democracy scores for the United Kingdom and Australia.[1] Polity does not score colonies, and so for India I give a brief description of the state of electoral and party politics in the colony.

In this chapter I establish that Great Britain's national identity focused on its island geography and naval power; its Protestant religious traditions and democratic political institutions; a commitment to spreading Protestantism, democracy, and "white civilization" through the Empire; and its imperial wars and wars to "save the continent" from various forms of tyranny. It had a military organization focused on developing small-unit cohesion through recruitment, training, rotation system, leadership, and esprit de corps. And it was an institutionalized democracy.

India's national identity was divided among competing narratives—secularist, Hindu, Muslim. Indians who identified with these different national narratives had different visions of India's politics, religion, and purpose in the world—different versions of the myths and symbols that made up Indian national identity. The Indian Army's military organization was very focused on small-unit cohesion; recruitment, training, and replacement policies were all designed to facilitate cohesion, and there was a strong tradition of esprit de corps and leadership. Finally, India was not a democracy, though it had some recent and limited experiences with elections. The British government retained final say over political and legal questions, meaning that elections were not truly open, competitive, or politically meaningful.

Australia's national identity emphasized its place in the British Empire, as well as its unique geography and democratic traditions, and its belief in white supremacy. Australians saw themselves as spreading white civilization to the far reaches of the globe. Military organization did not facilitate small-unit cohesion. Its recruitment and training were all conducted on a mass, national level, and replacement was generally individual rather than by entire unit. It was also a democracy in which white residents enjoyed substantial political equality. Non-white residents did face repression but were not represented in numbers in the military.

GREAT BRITAIN

At the beginning of World War II, Great Britain was made up of England, Scotland, Wales, and Northern Ireland. While about 50,000 citizens of Northern Ireland (and 70,000 from the Irish Republic)[2] volunteered for service with the British military during the war, conscription was restricted to England, Scotland, and Wales. For that reason, I exclude Northern Ireland from consideration in this chapter.

National Identity

British national identity was a layered identity, more than the sum of its English, Scottish, and Welsh parts.[3] Living together on an island archipelago, for the most part undisturbed by outsiders, these peoples developed what David Lowenthal calls the world's "most strongly defined sense of national identity."[4] Great Britain's status as both insulated island and global power informed its national identity, as did its history as one of the forerunners of liberal democracy. Democracy, Protestantism, an island topography, and a role as protector of Europe and manager of a great empire served as the myths and symbols that made up British national identity.

Homeland

Great Britain's geography is an important element of its people's self-understanding. Two major features shape British identity—its island status and the countryside. As an island, Great Britain developed its culture and identity without the constant threat of invasion present on the European continent. Though the British perceived threats from the other great powers—especially the Catholic great powers—war and conflict did not have the same deep impact as on the populations of continental powers. According to historian Henry Buckle, more than any other nation, the British had "worked out their civilization entirely by themselves."[5] Though the British were obviously influenced by other peoples in everything from religion to political philosophy, their isolation from Europe provided space for them to work out their own unique identity and to claim independence from outside influence.[6]

The sea thus served as an indicator of safety, a strong defense against any would-be aggressors. Moreover, the sea was also the foundation of Great Britain's power. Not only did it provide a defense of the homeland, but it also served as a means for Britain to expand its political and economic reach beyond its own shores. The Royal Navy ruled the waves, protecting British trade and defending its imperial holdings around the globe.[7] The navy held a place of honor in British society as defender of the island and its commercial way of life.

The countryside was another source of pride, though it could also serve to distinguish the ethnic groups on the island. While a single island, Great Britain's geography is diverse and divisions in topography correspond loosely with divisions of ethnic groups. The image of the British countryside described in poetry and literature was usually that of the English countryside, especially by the early twentieth century and the advent of automotive-based tourism.[8] Scottish nationalist literature emphasized the distinct landscape of the highlands, reinforcing the distinction between England and Scotland through topography. The countryside, hedge-lined lanes, and thatched roof cottages were primarily a product of England. More specifically, they were a product of English care and stewardship. It was not the wild, untamed moors that represented England but the carefully ordered country village, cared for by the landed aristocracy.[9] According to Lowenthal, this landscape represented four national traits: insularity (the island), artifice, stability, and order (the cared-for status of true English nature).[10]

It was not the topography itself that evoked feelings of Britishness but the social order and hierarchy it represented. The British aristocracy had popularized "country pursuits" as leisure in the late nineteenth century and continued to hold cultural influence through the interwar period (and beyond) after many other countries had overthrown their own aristocracy.[11] Moreover, Black points out that popular literature such as detective novels "often contrasted rugged English heroes, who relied on their fists, with foreign residents of London, who were generally presented in terms of supposedly undesirable physical characteristics, such as shifty looks and yellowish skin, and as using knives."[12] Cities were associated with foreigners, racial others, and physical and moral degradation.

British territory and landscape thus had a dual effect on British national identity. On the one hand, the island status drew the island's three ethnic groups together. It provided a sense of security as well as power in the form of the British Navy, which all groups depended on and in which they all served. But the Anglocentric nature of British identity shines through in the focus on the English countryside as representative of British identity. Thus, while the island geography drew the inhabitants together, topography could sometimes remind them of differences as well as commonalities.

Religious and Political Principles

When parliament passed the Act of Union in 1707, unifying Scotland with England and Wales, it did not so much combine three distinct nations as draw together a patchwork of identities.[13] Scots were as much divided between highlanders and lowlanders as were Scottish and English, and economic and religious divisions acted as crosscutting cleavages throughout the archipelago. In fact, as the political union continued to deepen over the next 200 years, these identities continued to develop alongside and within a British identity.[14]

Religion served as an important marker of British identity from the Reformation onward, even into World War II.[15] British identity in part developed out of a sense of defending the true faith of Protestantism against the tyranny and hypocrisy of Catholicism.[16] Religious divisions were reinforced by the Jacobite threat to the north. Living in and supported by France, these Catholic sympathizers and former members of the British Monarchy threatened the political and religious stability achieved through the Glorious Revolution of 1688. Throughout the eighteenth century, whenever Great Britain was weak the potential for invasion from France was high, and given the ties between the Jacobites and Scottish Catholics, the threat came from the north.

As the threat of invasion declined throughout the nineteenth century, so too did persecution of Catholics, with Catholic emancipation in 1829 allowing Catholics in Great Britain to serve in the government and paving the way for the franchise. However, Colley points out that while political restrictions were lifted, Great Britain remained a Protestant country.[17] Even after religion in general began to decline in the interwar years, Protestant symbols and culture remained an integral part of British identity, though less prominently than in earlier years.[18]

As political union developed, political issues and grievances became more British and less regional.[19] The home-rule movement in Ireland was particularly important in shaping a sense of British unity. Irish demands for political autonomy in the 1850s raised fears that the Scottish, Welsh, and English would demand the same. Some politicians suggested a "federal solution" to the problem of multiple nationalities, yet there was almost no support for this among the Scots and Welsh themselves.[20] In fact, as the franchise expanded to include Catholics and a greater proportion of the working class, all three ethnic groups began to feel themselves a part of a single entity. The expansion of the franchise and solidification of the Union were decided within a British context, rather than an English, Scottish, or Welsh one.[21] The same was not true of the Irish, where the potato famine and British racial attitudes continued to exclude Irish Catholics from a fully British identity.[22]

Politics did more than centralize government and draw ethnic groups together. The British believed that their long history of limitations on government and protection of individual liberty set them apart from their authoritarian neighbors across the channel. Liberty, the constitution, and the power of a representative parliament all informed British national identity.[23] Just as Protestantism set the British apart from their Catholic neighbors, so too did stable democracy and protection of liberty. This aspect of their identity sat uneasily with another important aspect—empire.

Purpose and Relations with the World

The British sense of place and purpose in the world was deeply informed by its empire as well as its commitment to democracy and Protestantism.[24] British imperialism began as an extension of British power into so-called "empty" territories: North America and later Australia and New Zealand. Importantly, these colonies included as subjects only the white settlers, considering the Indigenous peoples who did in fact live there as lesser others to be civilized and hopefully moved out of the way.

As white settlements grew (despite the loss of some territory to the United States), so too did nonsettler colonies in Africa, India, and Asia. This led to a dual understanding of British empire. The white settlements were considered an extension of Britain.[25] These settlements, as they grew in size and ability, would eventually become autonomous within an imperial framework.[26] Rather than risk another American-style revolution, Britain would ease the white colonies into self-rule as they asked for it but would continue to retain an imperial relationship with them; they were, after all, British. All while keeping non-white residents in second-class status.

Britishness did not extend so far as to include the non-white inhabitants of its colonies.[27] In these holdings, Britain drew material resources while providing "civilization" to the purportedly childlike inhabitants. As Powell notes: "There was also, though, a sense that Britain was providing the rest of the world with a moral lead, for example by taking the initiative in banning the slave trade and abolishing slavery within the empire, and more generally encouraging the spread of Christianity through the work of missionary societies and their 'civilizing' activities."[28] This view of the British project in the non-white colonies persisted until just prior to World War II. "In popular discourse as well as officially, the British understood their colonial relationships as fostering democracy and their nation as a benevolent, paternalistic imperial power."[29] Both white settlement and non-white colonialism were viewed as an expansion of British power and values throughout the world.[30] While the white dominions brought Protestantism and democracy to the "unpopulated" territories of Australia, New Zealand, and Canada, British civil servants and armed forces were "civilizing" the non-white residents of Africa, India, and Asia and making them ready for eventual self-rule.[31] White supremacy was a fundamental element of the empire and thus British self-identity.

The expansion of empire not only extended the reach of British identity around the world; it also helped to deepen it within Great Britain itself.[32] As the empire expanded, it opened new and important opportunities for participation in a broadly British project. "For some Scots, though, it was less the job and trading opportunities that empire provided, than the *idea* of empire that proved most compelling. If Britain's primary identity was to be an imperial one, then the English were put firmly and forever in their place, reduced to a component part of a much greater whole, exactly like the Scots, and no longer the people who ran virtually the whole show."[33] By increasing political opportunities and expanding the meaning of Britishness, the empire furthered the development of a British, rather than English, identity on the island of Great Britain.

Colonial wars were closely followed in the press, providing examples of British military prowess that posed little threat to the daily way of life.[34] According to Grainger, "empire promised a way not merely of not stagnating but also of remaining a great power in the world."[35] Welsh, English, and Scottish could all participate in this greater goal of civilizing the world and defending British power. Kumar points out that the elements of the empire were integral parts of life on the British Isles; it appeared in advertisements, music, novels, even the Boy Scouts.[36]

Britain's imperial identity did not go unchallenged.[37] There was a vocal minority that argued against expansion in general and particularly against the expansion of British identity outside of the islands of Great Britain. One leading voice for this group was G.K. Chesterton. Arguing for a particular form of patriotism, Chesterton insisted that the empire was too large a thing to feel patriotic about.[38] While conflicts such as the Boer War might draw support as a fight for fellow British—if not Britain—this was not "patriotism of the head and heart of Empire."[39] Chesterton sought to promote a limited, insular patriotism that focused on "little England," on the domestic characteristics and institutions that made England/Britain unique and to which loyalty was owed.[40] Other critics pointed to the enormous financial outlay and military risk garnered by imperial holdings as a reason they should be abandoned.[41]

After World War I, the empire expanded territorially but its ties to British identity weakened.[42] The white dominions increased their autonomy from London, eventually resulting in the Statute of Westminster and legal sovereignty for those territories. Colonial groups pushed publicly and in some cases violently for independence. Though the government continued to attempt to educate the British population about the goals of empire—political development and eventual independence—the British people grew ambivalent about their vast holdings.[43] The British public generally disliked independence agitators such as Gandhi, but there was a vocal minority who spoke out for independence and against empire.[44] "Britain's imperial relations across the globe subverted the framing of the war [WWII] as one being fought to secure freedom and democracy for both the country and the empire, and this was quite obvious to Britain's critics, both at home and abroad."[45] Though empire was a deeply rooted element of British national identity, its contradictions with democracy, freedom, and self-determination posed a problem in describing the goals of war against Germany and especially Japan.

War

Britain's wars can be broken into two categories—those for defense of Protestant, liberal Britain against Catholic, tyrannical France and those for the empire. The first group of wars spanned from Queen Anne's War in 1702 until the defeat of Napoleon at Waterloo in 1815 (though empire did lurk behind these wars as well).[46] These wars helped to define Britain over and above the constituent nations of England, Wales, and Scotland. According to Colley, the British nation "was an invention forged above all by war. Time and again, war with France brought Britons, whether they hailed from Wales or Scotland or England, into confrontation with an obviously hostile Other and encouraged them to define themselves collectively against it."[47] The threat of revolution or invasion reinforced Britain's Protestant and democratic identity. Additionally, Britain's victory over Catholic absolutism and Napoleon's hegemonic goals helped establish the idea within Great Britain that it was the defender of freedom and the independence of European countries as well as its own colonies.

The second group of wars were those fought for defense of the empire abroad. The largest of these, the Boer War, was also fought at a time of increasing popular press and thus received detailed and widely consumed coverage.[48] This war was a symbol of imperial unity. The Boer War was the first war in which the white dominions actively participated, and it was billed as an example of common interest and mutual support within the empire.[49] Yet it also brought early questioning of the empire. The conduct of the war and use of "concentration camps" combined with an initially poor military showing led some to wonder if defending the empire was moral.[50]

The Boer War as unifier and underminer of empire was soon surpassed by the experience of World War I. This war was not fought for self-defense but rather for a set of principles—defense of small nations and standing up to aggression.[51] Drawing together soldiers from around the empire, the fighting itself had the effect of drawing the empire into a single unit: "By suffering alongside Geordies and Brumies, Cockneys and Scousers, Micks, Jocks and Aussies, the Taffs became part of a new brotherhood: To become a soldier was to assume a new nationality."[52] Political leaders as well as soldiers saw the imperial war effort as a sign of imperial unity. Leo Amery, undersecretary during the war and later colonial secretary, wrote, "we fought the war as a united Commonwealth and Empire, and in the course of it achieved a greater measure of effective Imperial unity in its direction than statesmanship had ever contemplated before, or has achieved since."[53] World War I demonstrated what was possible for imperial cooperation and unity, but such cooperation could not be maintained in peacetime.

Yet the experience of WWI led a generation to question their political leaders and the value of patriotism. "Heroic" nationalism was tied to a war that decimated the people of Great Britain.[54] Moreover, the perception that political leaders had lied about German war crimes led a generation of British citizens to deep skepticism of such claims in the lead-up to this second war.[55]

Small-Unit Cohesion

The basic form of the regular army was established by 1881. This form placed primary emphasis on the regimental system: volunteers joined a specific regiment, trained with it, were sent abroad with it, and if necessary fought with it. This organizational system thus dictated recruitment, training, postings, replacement systems, and war planning. The system is a prime example of an organization dedicated to small-unit cohesion in the form of regimental honor and esprit de corps.

Recruitment was conducted on a primarily regimental basis.[56] Though some signed up for General Service, allowing themselves to be posted to the regiment chosen by the government, most recruits volunteered to join a particular regiment.[57] Ostensibly, regiments were meant to recruit from their associated territorial area, but in reality this often failed to be the case.[58] Generally, less than 50 percent of soldiers in a regiment were drawn from that unit's formal territory.[59] Units such as the Scottish Argyll and Sutherland Highlanders were thus made up of soldiers from all of Great Britain rather than just the indicated Scottish Highlands.

Once he joined a regiment, a volunteer immediately began training with that regiment.[60] Basic training had three purposes: "(i) the development of a soldierly spirit. (ii) The training of the body. (iii) The training in the use of rifle, bayonet, and spade."[61] To fulfill these goals, senior officers were given wide latitude in interpreting doctrine for training. Division commanders were essentially responsible for determining what sort of training was necessary to fulfill the demands of army doctrine.[62]

Training included not only military drills and physical activity but also education in the history of the regiment.[63] Regimental associations, made up of former regiment members, produced regimental histories that were issued to new recruits.[64] Officers were also expected to give lectures on the regiment's accomplishments as well as those of Great Britain. Furthermore, each regiment had its own set of traditions, rituals, symbols, and codes of conduct by which members were expected to abide. David French argues that, in many ways, the regimental system was based on a series of small "imagined communities" to which members owed allegiance and from which they drew their identities.[65]

Training also included an emphasis on hierarchy within the regiment. Officers held a special place. Not only were they seen as commanders and leaders in battle, they also served a paternal role within the regiment. Generally drawn from the middle and upper classes, officers ranked higher both socially and militarily than their soldiers. Nearly all had attended a public school, and all attended either Sandhurst or Woolwich. From these military colleges they were then posted to the regiment where they would spend the rest of their career.[66] There they were expected to lead their soldiers in battle as well as everyday life. Standing orders that regulated the minutiae of day-to-day living for soldiers encouraged officers to view their men as children who needed to be guided, supervised, and disciplined.[67] Regimental life was built, in part, around an officer corps that fulfilled a paternal role for its soldiers, a key element of Shils and Janowitz's argument that attentive officers who care for soldiers' basic needs contribute to small-unit cohesion.[68]

According to Brian Bond, the regimental system was the most important facet of the British Army's character between WWI and WWII.[69] It was so ingrained in British Army culture and organization that the prohibition against cross-posting soldiers to another regiment was observed as late as 1943, when casualties and training needs finally overrode the desire to keep up esprit de corps and regimental honor.[70]

Britain's military organization was focused on the creation and maintenance of small-unit cohesion. Soldiers were recruited into specific units, trained with those units, and were sent abroad and replaced as a unit rather than as individuals. The organization focused on officer development, training them to look on their men as in need of a father figure to guide them. Additionally, the British built a strong sense of esprit de corps in their units through the use of history and traditions. That esprit de corps was intended to make soldiers loyal to their specific unit and encouraged them to see themselves as upholding their unit's honor or even increasing that honor in battle.

Democracy

Polity V gives the United Kingdom (England, Wales, Scotland, and Northern Ireland) a score of 10 for 1939 (its score since 1922), the year the war began.[71] At the end of 1939 the United Kingdom had significant experience in competitive and open executive recruitment for office, constraints on the chief executive, and competitive political participation. Literacy was high in Great Britain, with 97 percent of adults literate by the turn of the twentieth century.[72]

INDIA

National Identity

India had multiple national identities at the beginning of WWII. Though many Indians supported some form of independence from Great Britain, there were competing narratives of what made India a unique nation. Secular nationalists—mostly in Congress, though there were also a number of more radical groups—competed with Hindu nationalists and Muslim nationalists regarding the myths and symbols that made up Indian national identity.

Congress Secularist Nationalism

Partha Chaterjee argues that Indian nationalism in this period had two parts: the political, independence-focused part (outward) and the "spiritual" (inner) part that focused on cultural identity distinct from the West.[73] I include discussions of both elements of Indian national identity below.

All-India nationalism took off at the end of the nineteenth century with the founding of several all-India associations, the most prominent and long-lasting of which was the All-India Congress Party (Congress). Established in 1885 by English-educated Indians and former (British) members of the Indian government, Congress's goal was to increase representation of Indians in the British-run government. It was not initially an independence movement: "the progress of the Congress movement from 1885 to 1905 was one even march based on a firm faith in British rule and the justice of the English nation."[74] Increasing unemployment, especially among the educated, and the influence of nationalist movements abroad led Indian nationalists, including Congress members, to take a harder stance with the British government. In 1906 it declared dominion status as its goal.[75] By the 1920s Congress was organizing mass noncooperation and boycott movements. More extreme groups such as the Hindustan Socialist Republican Army conducted acts of violent resistance to British occupation but also engaged with Congress on many issues.[76] Congress was the best organized and best funded of all the nationalist organizations, but it was not the only voice, nor did it speak as one at all times.[77]

HOMELAND

What are now three countries—Pakistan, India, and Bangladesh—were all part of British India under colonialism. Congress's focus was on uniting all of India and creating a single independent state out of the entire region. On its founding Congress sought to incorporate the numerous regional political associations already in existence in order to develop an all-India organization. It drew on the political experience and groundwork of these organizations, even borrowing their leaders for its own organization.[78]

Uneven development across India meant that Congress was not geographically representative. The regions closer to the coast, and those that received greater financial resources from the British, had more educated populations and thus were more heavily represented in political organizations including Congress.[79] Moreover, Muslims were concentrated in a few geographic areas—Punjab, Bengal, and Bombay, primarily—reinforcing geographic differences within the organization.[80] Congress worked hard to promote a version of Indian nationalism that included all the territory of colonial India, despite those borders being a product of the colonialism they wished to shed. Homeland was thus not an agreed-on concept in India, though secular and Hindu nationalists considered all of British India to be India.

POLITICAL AND RELIGIOUS PRINCIPLES
Congress envisioned a socialist democracy, free from the problems of Western capitalism. Congress nationalism contained a number of Marxist or leftist elements, especially on international issues. Socialists rejected British imperialism, something with which Indians could identify. After the Bolshevik revolution, British opposition to the new communist government in the Soviet Union cemented Indian support for international communism.[81] Moreover, Indian nationalists claimed practices of democratic governance and rule by consensus as part of their political history.[82] Democracy did not belong to the West alone.

Gandhi's rejection of mechanization and capitalist industrialization in favor of traditional crafts such as handwoven cloth linked Indian civilization to trends in international socialism. Socialism also provided a set of critiques of Western civilization on which nationalists could draw. Gandhi argued that the British domination of India was not simply a single fault but a symptom of the fundamentally flawed civilization in which they participated.[83]

As the world drew closer to war in the late 1930s, Indian nationalists were strongly opposed to fascism in Europe and Japan.[84] They accused the British of colluding with the fascists to destroy the socialist Soviet Union and were highly critical of British foreign policy.[85] Yet opposition to fascism did not translate to support for the British war. Congress tried to organize a noncooperation movement against the British war efforts in 1938, taking the slogan "na ek pai, na ek bhai" (not a pie, not a man) and urging the people not to cooperate with or support the army. The movement failed to gain much traction, not least because the army was a sure means of income for a peasantry facing high unemployment and rising cost of living.[86]

Congress (and other, more revolutionary groups) took a position of secularism, arguing for the unity of all Indians no matter their creed.[87] Islam, Hinduism, Sikhism, and other religions were all represented among Congress's members and leadership. Until the late 1920s those Muslims that were politically active found it possible to work with Congress for political independence for all of India.[88] Congress declared that divisions between religions in India were a product of British rule and argued that the British encouraged Muslims to fear their position in India and to put up roadblocks to independence in the form of separate representation to avoid granting independence.[89]

Congress leadership, as well as groups such as Naujawan Bharag Sabha (NJBS)—a radical student organization—insisted that secularism was the only way to move forward as a unified and independent nation.[90] Jawaharlal Nehru, who also had ties to NJBS, was especially committed to secularism, eschewing most references to religion in rhetoric and policy.[91] The communal riots and disagreements of the late 1920s seemingly affirmed this need and led them to reject any political arrangements based on religion. Congress secularists attempted to come to an arrangement with Muslim political groups in the early 1920s, but their refusal to allow separate electorates and reserved districts for Muslims meant that no fundamental agreement was possible. After a few failed negotiations, Congress began ignoring the religious question, insisting that it was a product of British divide-and-rule tactics and would disappear with independence.[92]

Nehru and Gandhi took very different views of how secularism should inform Congress's and thus India's identity. For Nehru, religion had no place in politics at all. It was divisive and was fundamentally distinct from politics. He even doubted the capacity of Muslim organizations to operate in politics, as they were religious bodies: "the time has gone when religious groups as such can take part in the political and economic struggle. That may have been so in medieval times: it is inconceivable today."[93] Religion was not a legitimate political identification, and groups that organized themselves along those lines were unfit for political activity.

Gandhi, on the other hand, was much more accommodating to Muslim concerns, especially before the communal riots. He also attributed divisions within India to British policies but acknowledged that they did exist. In 1910 he wrote: "There is mutual distrust between the two communities. The Mohammedans, therefore ask for certain concessions from Lord Morley. Why should the Hindus oppose this? If the Hindus desisted, the English would notice it, the Mohammedans would gradually begin to trust the Hindus, and brotherliness would be the outcome."[94] Gandhi personally managed the cooperation between Muslims and Congress on the Khilafat movement, which sought to maintain the caliphate in Turkey after its defeat in WWI. Gandhi saw the movement as an opportunity for Hindu-Muslim unity and worked closely with leaders of the Muslim community to organize civil disobedience and boycotts. However, the cooperation could not survive his imprisonment.[95]

Despite Congress's status as a secular organization and the leadership's commitment to that status, Congress was not free from religious rhetoric and influence (nor were other avowedly secular groups). As we will see below, Hindu language and imagery suffused much of Congress's rhetoric and even influenced its boycotts and protests. Metcalf and Metcalf argue that "nevertheless, Gandhi's entire manner, dress, and vocabulary were suffused with Hinduism."[96] Formally, though, Congress put forward an image of India as bearer of an ancient and great civilization with a socialist ethos, unified despite its variety of religions.

PURPOSE AND RELATIONS WITH THE WORLD
Indian nationalists drew on the findings of a growing Indian historiography as a narrative of the Indian nation. Gandhi was especially keen to distinguish between Western civilization, which he viewed as shallow and corrupt, and Indian civilization, which he believed emphasized duty and morality.[97] In one of his most influential tracts, published in 1910 while he was working in South Africa, Gandhi wrote: "It is not due to any particular fault of the English people [that they are fickle in their politics], but the condition is due to modern civilization. It is a civilization in name only. Under it the nations of Europe are becoming degraded and ruined day by day."[98] Modern Western civilization, with its emphasis on wealth and technology, did not better its members. Instead, it corrupted them, encouraging the people in their vices and indulgences.[99]

In contrast, Indian civilization developed in its members a sense of self-control and duty toward society. He described that civilization:

> I believe that the civilization India has evolved is not to be beaten in the world. Nothing can equal the seeds sown by our ancestors ... Civilization is that mode of conduct which points out to man the path of duty. Performance of Duty and observance of morality are convertible terms. To observe morality is to attain mastery of our mind and our passions. So doing, we know ourselves. The Gujarati equivalent for civilization means "good conduct." ... Our ancestors, therefore, set limits to our indulgences.[100]

Gandhi eschewed Western or modern civilization, along with the educational and technological innovations that made it possible. He blamed India's economic woes on trains and factories[101] and its internal divisions on

British policy and law (and the Indian lawyers who perpetuated them).[102] He believed that the solution to both the civilizational problem and the problem of British political domination was to return to Indian civilization's teachings of self-control and duty. Gandhi's arguments for nonviolent soul power, noncooperation, and boycotts all stemmed from his understanding of Indian heritage and civilizational teaching. Secularist Indians saw their nation as one that could bring about a reconciliation between East and West, drawing on the best of both cultures to offer a more human society to the world.[103]

WAR

Congress viewed the Indian Army and its participation in British colonial wars as both a tool of colonialism and a potential tool for nation-building. Before and during the First World War, Congress supported the British war effort and encouraged Indians to join the military and fight for the British. The British promised political changes after the war, and Congress had high hopes that support in the war would lead to British gratitude and increased autonomy for India.[104] When the war ended and no political reforms were forthcoming, Congress abandoned most of its interest in the military.

The one issue in which Congress maintained interest was the officer class and the move to allow Indians to serve as senior officers. Congress believed that as Indians acquired more of a stake in the military their loyalty to India as a political unit would increase. Additionally, Congress leaders were looking ahead to independence, when they would need an officer corps to run an independent Indian Army.[105]

Congress leaders learned their lesson from WWI and refused to pledge their support to the British as war loomed in the late 1930s. They demanded actual political changes and a concrete timeline for independence rather than broad promises to be implemented after the war.[106] Congress viewed British wars as a chance to leverage political reforms and increase India's independence. Whereas in WWI they attempted full support in order to gain those changes, before WWII Congress planned to withhold support until Great Britain met its demands. War was not so much a defining element of Congress nationalism as a tool.

Hindu Nationalism

Despite Congress's domination of the political landscape, there were other political organizations within India vying for control over the nationalist narrative. Some of these organizations put forward a view of India as a distinctly Hindu nation in which Hindu practices should be privileged and protected. These organizations had enough popular support to play spoiler in negotiations between Congress and Muslim groups, and in negotiations with the British government. The use of Hindu language and symbols in nationalist rhetoric was widespread, even among Congress leaders who promoted a secular vision of the Indian nation.

Hindu nationalists differed from Congress primarily in their view of the importance of Hinduism in Indian national identity and thus their exclusion of Muslims from that identity. Because Hindu nationalists did not differ from Congress or secularist nationalism in their views of Indian homeland, purpose and relations with the world, or war, I outline here only their unique views on Hinduism and Indian national identity, views that did appear in some Congress and even revolutionary rhetoric.

POLITICAL AND RELIGIOUS PRINCIPLES

At the end of the nineteenth century, at the same time that all-India nationalism was picking up momentum, a movement for Hindu revival also emerged. A number of associations formed with the goal of purifying Hindu practice or improving society along Hindu lines. Organizations such as Arya Samaj (a Hindu social improvement association) sought to increase and purify the practice of Hinduism throughout society. They agitated for cow protection and banning cow slaughter (practiced by Muslims during some of their religious festivals), and urged a revival of folk crafts such as cloth weaving that had ancient religious associations.[107] The cloth weaving revival became a prominent part of Congress's noncooperation movement in the 1930s.

The issues of cow protection and cloth weaving were brought together during the cloth boycott in the 1930s. Congress leaders urged the boycott of all cloth from England and its replacement with handwoven cloth made in the traditional Indian manner. This boycott was one of the first mass movements of Indian nationalism and was meant to demonstrate a rejection of the British economy and industrialization. The campaign drew heavily on religious symbols. First, numerous Congress leaders accused English cloth makers of using cows in the production of cloth, making it religiously as well as politically taboo. This was a serious accusation, as the British claimed that concern over cow fat in gun cartridges had sparked the 1857 mutiny.[108] Nationalist leaders argued that using English cloth symbolized a rejection of the Indian nation through economic support of Britain and the violation of Hindu religious practice.[109] Second, local Congress leaders encouraged the population to take an oath swearing not to purchase or use English cloth; the oath was administered at Hindu temples and religious ceremonies, effectively excluding the Muslim population from this particular form of participation in Indian nationalism. Participation in the boycott was enforced through traditional caste sanctions, thus requiring unique sanctions for Muslims. These sanctions turned out to be sometimes violent picketing of Muslim stores that sold English cloth.[110]

Arya Samaj also urged the elimination of caste distinctions. Caste uplift was said to be a duty of the higher castes toward the "depressed castes" that brought the untouchables fully into Hinduism. Arya Samaj thus worked to eliminate social distinctions based on class in order to increase Hindu unity, a goal Gandhi supported as well.[111] Social changes such as the elimination of caste were explicitly promoted in order to homogenize and consolidate the Hindu population of India and—purposefully or not—excluded the Muslim population.[112]

In addition to revivalist points of emphasis such as cow protection, cloth, and untouchable uplift, Hindu religious symbols were also an important part of some nationalist rhetoric. Nationalists often used itinerant teachers to help spread the nationalist message, thereby associating nationalism with these Hindu holy men.[113] Hindu festivals and holy days were often used as platforms for nationalist speechmaking—both Hindu-nationalist and Congress. Even if Congress leaders did not use Hindu language or symbols, their use of the religious festival associated their message with Hinduism.[114] Congress leaders also attempted to use Islamic festivals and events to spread their message, but their efforts in that direction were half-hearted and ineffective.[115]

There were also more explicit uses of Hindu religious imagery in some nationalist rhetoric. Hindu Mahasabha, an explicitly communal political organization, emphasized Hinduism as the defining element of Indian national identity.[116] But even Congress leaders, especially at the local level, used Hinduism as shorthand for the Indian nation. There are numerous speeches and pamphlets that analogized India's struggle against the British with the battle between Ram and Ravan.[117] Nationalist leaders—especially Gandhi—were often deified using traditional Hindu language about the gods.[118]

Kama Maclean, in her analysis of nationalist posters during the interwar period, points to numerous religious themes that inform their portrayal of nationalist martyrs and leaders, even those who adhered to a secular notion of the nation.[119] She points to posters portraying executed nationalist Bhagat Singh in the pose of Krishna, drawing on Hindu themes. Portrayals of nationalist leaders as crucified (Christian themes) and as having been beheaded (Sikh themes) drew on a range of Indian religious traditions.[120] Muslim religious imagery does not, however, appear to have been common in these posters.

Finally, Gandhi himself used religious language, even while proclaiming India a secular nation that included both Hindu and Muslim. In his *Hind Swaraj* essay, he wrote: "The British Government in India constitutes a struggle between the Modern Civilization, which is the Kingdom of Satan, and the Ancient Civilization, which is the Kingdom of God. The one is the God of War, the other is the God of Love."[121] Gandhi argued that secularism or religious tolerance was a fundamentally Hindu principle.[122] According to Gandhi, Hinduism had enough room within it for both Christianity and Islam.[123] Members of those religions did not necessarily agree.

Hindu revivalism, though ostensibly religious, required serious social changes and influenced the rhetoric and tactics of Congress's secular nationalism. Hindu nationalists such as the Hindu Mahasabha explicitly stated that the Indian nation was a Hindu one. Even Congress drew on Hinduism to illustrate and help spread its message. Using the language of Hinduism helped Congress to spread Indian nationalism beyond its urban elitist roots but also alienated the non-Hindu members of Indian society.

Muslim Nationalism

Unable to identify with the religious symbolism that infused much of Indian nationalism and fearing for their political and religious rights as a minority in a Hindu-majority state, Muslim nationalists moved slowly from supporting the British regime to seeking independence in a separate state. Though there were periods of cooperation with Congress, and some individual Muslims were important Congress members, the Muslim population of India generally proceeded on a parallel track toward independence—from cooperation with the British, to petitioning the government for rights, to demands for outright independence. But Congress's association with Hinduism and inability to accommodate Muslim fears, and Muslims' identification with the pan-Muslim world, meant that Indian Muslims did not view themselves as part of a unified Indian nation.

HOMELAND

After the British established their rule in India, Muslims often had less opportunity for education and economic development.[124] As a result, in those locations where Muslims did retain some political power, they were loath to give it up to the majority Hindu population.

Because of the concentration of Muslims in three main regions—Punjab, Bengal, and Bombay—Muslim nationalist groups were more comfortable with changing India's borders than were secular and Hindu nationalists. When in the 1930s the Muslim League gave up trying to negotiate a political settlement with Congress, its leaders began to suggest a territorial solution in the form of a separate Muslim state—Pakistan. It adopted the Pakistan Resolution in March 1940.[125]

In addition to the demographic appeal of a distinct Muslim state, Muslim nationalists were also informed by loyalty to the extraterritorial Ummah—the ideal Islamic state, believed to have ended with the defeat of Turkey and breakup of the caliphate.[126] The height of Muslim-Hindu cooperation in India was during the early 1920s, when both groups lobbied the British urging them not to participate in dismantling the caliphate.[127] Blaming the British for breaking up the Ottoman Empire, Muslims declared themselves to be anti-British and in favor of independence as well. Yet they were not committed to a territorially united India but rather to a non-British state for Muslims.

POLITICAL AND RELIGIOUS PRINCIPLES

Early in the Imperial period Muslims saw the British as their protection against the Hindu majority. When the main Muslim nationalist organization—the All-India Muslim League—was formed in 1906, its first object was "to promote among Muslims of India feelings of loyalty to the British government and to remove any misconceptions that might arise as to the intention of the Government with regard to any of its measures."[128] The Muslim League sought separate representation of Muslims in any representative body and guarantees of their rights (such as cow slaughter, which the British protected).

Bengal was partitioned in 1905, making one state that had an even demographic split between Hindus and Muslims into one Hindu-majority state and one Muslim-majority state. Hindus objected vociferously.[129] When the partition was annulled in 1911, Muslims took this as a sign that the British were no longer going to protect them in return for their loyalty.[130] Britain helping to dismantle the Ottoman Caliphate solidified that feeling.[131]

As the elites continued to clash on political arrangements, both Hindus and Muslims grew suspicious of the other's use of religious figures and symbolism in their political rhetoric. The Hindu nationalist rhetoric described above deeply unsettled and excluded Muslims. But the Khilafat movement also used religious figures and symbols to further their own political goals.

Just as Congress used Hindu itinerant priests to spread the word about Congress goals and mass actions, the Khilafat movement used Islamic imams and religious leaders to urge popular support of noncooperation.[132] Additionally, the movement focused heavily on the religious importance of the caliphate and the sultan's role as head of Islam in the world in their arguments to the British. In an effort to appeal to Britain's self-understanding as religiously tolerant, the Khilafat movement alienated Hindus who feared Muslims might impose Islam in India.[133]

Once the British had helped dismantle the Ottoman Empire and Ataturk had divested the sultan of his political power in 1924, the Khilafat movement lost its main purpose.[134] Relations with Congress grew increasingly strained, and political accommodation looked less and less likely. The Khilafat leaders fell back on goals of communal defense.[135] Growing fear of Hindu nationalism and disappointment with the British government led Muslim leaders to search for a new political solution.

By the time WWII began, Congress and the Muslim League were in direct conflict. When Congress ministers resigned their government posts after the viceroy's declaration of war, the Muslim League declared its support for the British and celebrated the Congress's loss of power, hoping to increase their political value to the British. The head of the Muslim League declared that day the "Day of Deliverance."[136]

PURPOSE AND RELATIONS WITH THE WORLD

As noted above, Muslims considered themselves to be part of a nonterritorial political community: the Ummah. Muslims in India were influenced by the Wahabi movement that began in Saudi Arabia but spread throughout the Islamic world.[137] They sought to live a pure version of Islam and to promote its growth in the world. For that reason, some Muslims refused to participate in WWI when Turkey fought England, seeing that fight as a "jihad" to defend Islam against the infidel.[138] When the caliphate fell, pan-Islamism grew even more powerful in India, drawing Muslim loyalties away from the Indian nation and toward their coreligionists.[139]

Muslims in India did not necessarily want to break away from the Hindu regions but were insistent that they have communally based representation to protect their interests against Hindu encroachment.[140] Secular and Hindu nationalists lambasted them as pro-British and anti-national, but Muslims feared the Hindu content of Indian national identity and that increased self-government would lead to encroachment on their abilities to practice their religion, such as restrictions on cow slaughter. Muslims in India thus saw themselves as threatened by the Hindu members of the Indian nation and as part of a nonterritorial community of Islam. This sense of place in the world, as alienated from their neighbors and members of a different political-religious community, mitigated against Muslim identification with Hindu or even secular Indian nationalism.

WAR

Muslims in India were even more skeptical of India's imperial wars than were secular and Hindu nationalists. Muslims had been very concerned about participating in WWI, as they would be fighting against their coreligionists from Turkey, and even fighting against the leader of the Islamic Caliphate.[141] After the war, Muslims also noted Britain's failure to follow through on its promises of political reform and were very skeptical of similar promises in the lead-up to WWII.[142] Despite the Muslim League's offer to support the British war effort, other Muslim groups reminded their coreligionists of Britain's failed promises and attack on the caliphate. Though many Muslims served in the Indian Army, India's recent wars offered no source of glory or loyalty to India's Muslim population.

Sikh Identity

Sikhs made up a small but important minority in interwar India, and especially in the Indian Army. Long considered by the British to be the most loyal of India's religious or ethnic groups, Sikhs actually played an important part in Arya Samaj, the Indian National Congress, and several moments of political rebellion in Punjab.[143] Kate Imy argues persuasively that Sikh identity was fluid throughout this period, allowing Sikhs at times to identify with Hindu Indians and at other times to distinguish themselves even as British officials worked hard to establish a unique Sikh identity loyal to the Raj.[144] There was not substantial support for Sikh political autonomy during this period, so, while noting the Sikh identity of Indian soldiers where possible, I do not consider the Sikh community as a distinct national identity. Rather, Sikh myths and symbols further complicate the range of Indian national myths and symbols operating at the time.

Small-Unit Cohesion

The Indian Army began as a system of mercenary forces hired by the East Indian Trading Company to fight small local wars. By the 1930s it had developed into a large volunteer defense force trained and officered primarily by British officers and organized to discourage caste and communal competition and develop strong small-unit cohesion. The Indian Army drew on the traditions of the British Army, as well as the theory of so-called martial classes for its organization and recruitment, and except for the two world wars, was generally used in border warfare and for local security.

After London took control of India's government, the Army was reorganized into a centralized, unitary force. Organization, discipline, training, and tactics were generally left to the military to determine, though civilians controlled the size of the military budget and overall expenditures.[145] The British government determined strategy, and the Indian Army functioned as an arm of the British Army, units of which were also stationed in India. Together the Indian Army and the British Army units stationed there made up the Army of India.[146]

The Indian Army retained the Company's focus on recruiting from particular castes and ethnicities that were believed to be "martial" or to have a long-standing warrior tradition.[147] A consistent feature of recruitment was that it systematically drew on the least politically developed classes and regions.[148] Initially units were recruited territorially but the focus on martial classes eventually eliminated any real territorial connections and focused instead on recruiting the specific castes and ethnicities believed to be most loyal and warlike.[149] Individuals were recruited into specific units composed of soldiers from similar caste and ethnic backgrounds, and stayed with those units for the entirety of their enlistment period.[150] Caste recruitment was dropped in favor of mass recruitment (though still into specific units) during both WWI and WWII.[151]

Class composition was the driving force behind the organization of regiments, battalions, and companies.[152] Muslims and Hindus were mixed at the battalion level but kept separate at the company level with their own food and religious arrangements.[153] This discouraged intercommunal or interclass competition while also providing for different religiously mandated living requirements. It also aided in communication, keeping soldiers of the same language and dialect together, aiding in small-unit cohesion. This organization also had the effect of reifying communal distinctions and preventing cross-communal socialization within the army.[154]

Until the 1920s the Indian Army had a two-tiered system of officers. Every company was commanded by a King's Commissioned Officer (KCO), usually a Sandhurst Military College graduate, and always a British soldier—never an Indian. The rest of the commissioned-officer positions were then filled by Viceroy's Commissioned Officers (VCO) who were Indian. An

Indian soldier could never command a British soldier and could never be the final authority for the unit.[155] The British commanding officer and his Indian subordinate officers often entered their unit together as young men and fought together over many years. The goal was to develop strong working relationships and close ties to the soldiers they commanded. KCOs and VCOs alike were trained in a strong sense of paternalism toward their soldiers, and the soldiers were trained to look up to the officers as they would a father figure. This was believed to encourage strong comradeship and small-unit cohesion.[156] It also built racial inequality into the organization of the Indian Army.

Political changes in India as well as the army's performance in World War I gave military officials reason to reconsider this system. The Indian Army units that fought in the trenches on the Western Front broke ranks and fled in a way Indian Army leaders had never seen.[157] Neighboring British units did not desert their own trenches, leading officers to consider why their prized Indian units collapsed. One possible explanation put forward was that by training Indian VCOs and soldiers to rely so heavily on their British officers, the army had in fact trained all initiative and leadership out of them.[158] Dependence on British leadership might be feasible when fighting lightly armed tribes in Afghanistan, but mass warfare on the scale of World War I required initiative and leadership from all levels of the troops.

The army decided to gradually introduce Indian commanding officers into the military by allowing a certain number of candidates to attend Sandhurst in England, dedicating particular units to be led by Indian officers, and eventually building a military training college in India.[159] Several problems arose once Indianization began. First, Indian cadets faced racist and classist discrimination at Sandhurst and then again as junior officers in the Indian Army; officer clubs were segregated, and the Indian officers were limited to specific "Indianized" units.[160] Second, the officers' role as members of a British military force as well as their own caste and class affiliations alienated them from socializing with other Indians in the army or the area in which they were stationed.[161] Finally, education focused on developing "masculine" men by merging the "martial race" thinking of the Indian Army with British "muscular Christianity," reinforcing racialized gender norms and the imperial racial hierarchy.[162]

The process of Indianization was still in its early stages when WWII began, leaving some units with Indian officers and most with British—as the war progressed and mass recruitment emerged, Indian officers became commonplace, though training did not necessarily adjust to that fact. Up until the beginning of the war, the British continued to view the army as a tool for defending against small, armed tribes and policing the unruly domestic scene.[163] It was trained for those purposes and not for fighting a mass war, though the doctrine it adopted for the war was well matched to the new environment.[164]

The Indian Army was organized with small-unit cohesion in mind. Soldiers were recruited into their unit, trained with it, and fought with it for most of their careers. Both officers and soldiers were trained in the importance of the officer-man relationship and saw the British officer as a benevolent father figure. The martial-class mythology was intended in part to give soldiers an identity to hold to and group whose honor they needed to uphold. There was a strong focus on esprit de corps in the Indian Army, in addition to organizational features encouraging small-unit cohesion.

Democracy

India was not an independent state at the beginning of World War II and thus is not scored by Polity V. Beginning in 1909 the British allowed for some Indian participation in government through legislative councils. Some council members were elected by direct vote (though the franchise was limited by property qualifications) and some by special interest groups.[165] In 1935 the Government of India Act expanded the franchise somewhat and granted more responsibilities to elected Indian officials.[166] Yet the British still retained control over foreign and defense policies, and the decisions and recommendations of elected ministers could be rejected by the British viceroy. While educational achievement varied across India, in 1941 only 35 percent of school-aged children were enrolled in school.[167] Thus educational attainment and practice in self-governance were both low in British-ruled India, suggesting that the democracy argument (which points to individual education) should not operate.

AUSTRALIA

National Identity

Its unique geography—distant from its British cousins and environmentally inhospitable within—as well as its ties to democratic Britain and the imperial project informed Australian national identity. Though Australia formed a self-governing federation in 1901, it maintained close ties to Great Britain and depended on it for military protection. It was also willing to aid its mother country whenever necessary, as demonstrated by its participation in multiple imperial wars. Australia considered itself to be a unique and important part of the British Empire.

Homeland

Australia's unique geography informed its national identity in two ways—the internal struggle with the land and its difference from Britain, and the external isolation from Europe and exposure to Asia. The nature of Australia's landscape fed nationalist imaginations from an early period, but the idealization of the bush and the bush life began in earnest in the 1890s when economic depression drove many to the cities that would remain the centers of Australian population from that time forward.[168] As drought and depression drove people off the land, writers and artists began to idealize the bush as something distinctively Australian. The weekly publication *The Bulletin*, first published in 1880, was a self-consciously nationalist publication that sought out and printed bush-ballad poetry, stories of the outback, and comic portrayals of city capitalists and aristocrats.[169]

The bush legend of independent workers traveling the land and supporting their mates did not actually describe the lives of most Australians, but it provided a sense of identity—something that distinguished the Australian from his British cousins and fed a strong sense of superiority in all Australian military endeavors. Mateship would become especially important in soldiers' attitudes about themselves and their fellow fighters.[170] Mateship had deep roots in the concept of the rugged Australian fighting for survival in the harsh outback.

While the exotic landscape and the mythic culture it encouraged provided a means for Australians to distinguish themselves from the British, Australia's isolated global location kept it firmly tied to the metropole. As discussed below, the racial "others" that were perceived as surrounding and potentially threatening Australia led it to deepen its dependence on Britain's military protection. Even while Australia feared that Britain would not be able to defend it against a Japanese invasion, it continued to make policy on the basis of imperial defense plans such as the Singapore strategy and even to tighten its links with the empire.[171]

Political and Religious Principles

Westminster-style parliament and ties to the monarchy replicated the British political system in Australia.[172] Australians saw themselves as enacting Britain's civilizing mission across the world. They were actively aiding Britain in the spread of democracy and civilization to the far reaches of the globe. As late as 1935 Prime Minister Menzies (who would be prime minister when war broke out) spoke to the association of Imperial Parliaments, pointing to the British roots of Australia's political system:

> I am confident that our parliamentary system is going to see it through. And why? Because its roots are deeply set in the history and character of the British people. In those countries where it has fallen, a parliament was adopted as the embodiment of an attractive theory. As a fully-grown tree it was carefully transplanted and watered and cared for. And at the first real blast of the storm it fell. Will our parliaments survive? I believe that they will. The growth of parliament is in truth the growth of the British people; self-government here is no academic theory, but the dynamic power moving through 800 years of national history.[173]

Australians saw themselves as a deeply democratic people because they shared Britain's democratic history. Their own democratic institutions and practices were rooted in their own national history. In some ways the Australians were even ahead of their British cousins in democratization; they extended the vote to all white women of voting age in 1902, sixteen years before Great Britain.

But democracy in Australia did not extend to all members of Australian society. While white women achieved the vote in 1902, Australia's aboriginal population would have to wait another sixty years.[174] As with many nations, the development of Australian nationalism cannot be fully understood without a close look at the role of race. Beginning with the first settlement at Sydney Harbour by transported convicts and soldiers, whiteness was one of the major defining features of Australian identity. While the first settlers may have been expelled from their own white civilization, they were still bearers of it in this new land, populated as it was by a people the Europeans believed incapable of civilization.[175] Violent encounters with the aboriginal population—usually with the Aboriginals bearing the worst of it—reinforced the distinction between white Australians and the continent's original inhabitants.[176] The Aboriginals remained until very recently an "other" against which the Australian nation could define itself.[177]

Racial differences served as a source of white-Australian unity both politically and in religion. When Britain sent convicts and colonists to Australia they brought with them their religious beliefs and practices. A number of Christian denominations sent priests, pastors, and missionaries to minister to the colonists and to convert the aboriginal population. Perhaps because they saw themselves as more similar to one another than to the non-white peoples of the region, differences in denomination were not as salient in Australia as in Great Britain.

The one point at which religious differences became contentious was during the First World War. As the Catholic population in Australia was mainly of Irish descent and led by Irish priests, they were very concerned with the state of Catholics in Ireland. Skeptical of the war from the start, Catholics in Australia were appalled by the 1916 Easter Rising and subsequent British repression. The Bishop of Melbourne, Bishop Mannix, voiced his opposition to conscription and began arguing against supporting Britain in the war.[178] This move helped prevent conscription and reignited some sectarian tensions.

Purpose and Relations with the World

Australians believed they played an important role in Britain's imperial mission. They were spreading democracy and white civilization to a foreign land and planting their seeds in Asia. As such, it was vital to Australian national identity that the country maintain its imperial ties with Britain.

The place of Australia in the British Empire is complex and its legal status debated among scholars.[179] However, there is no question that Australia's imperial membership was an important aspect of its national identity. Australian identity was multilayered, including loyalty to state, nation-state, and empire.[180] The development of a distinct Australian identity that did not do away with an imperial identity owes much to the racial and geographic issues discussed above.[181] Though Australians came from Ireland, Scotland, Wales and England as settlers and as convicts, they had in common a language and political culture as well as a race. Whether having left their own country through choice or by force, settlers recognized their common whiteness and belief in their civilization over and against the aboriginals and their Asian neighbors.[182]

As noted above, Australians modeled their political institutions on Great Britain and saw themselves as carrying on the long tradition of British democracy. But self-government did not require cutting ties with Great Britain or weakening the empire. After WWI some dominions began to reach for greater power within the British Empire as well as greater independence from it. The Irish Free State was formed in 1921 and, along with South Africa and Canada, insisted on further defining dominion status, rights, and obligations. Australia and New Zealand participated in these discussions only grudgingly, doing nothing to encourage further distinctions between dominions and Great Britain.[183] Australians were content to maintain their status as a dominion and even made moves to strengthen their ties with Great Britain through increased immigration and military dependence.[184]

The Statute of Westminster formalized the agreements reached by the dominions and Britain during the 1930s, increasing the dominions' powers to act independently of Britain. Yet Australia purposefully put off the application of that statute to its own relations with Britain, preferring to leave it unchanged. It was not until after WWII had been ongoing for over two years that the parliament finally adopted the statute, and then with little fanfare and only to allow necessary changes to maritime laws.[185]

The importance of race in Australian national identity became more clearly defined with the rise in Asian immigration to the continent. The mostly Chinese temporary laborers that began to arrive with the onset of the gold rush led to a spate of anti-Asian legislation from the Australian colonies. Exclusion of and discrimination against the Chinese became a point of contention with London but one of the first areas of major cooperation between the colonies.[186]

In 1895 Japan began to expand into China, giving rise to a threat to Australia's white identity in the Australian mind. The Japanese now assumed a position of racial and national "other" that they would hold for the next fifty years. The policy of "White Australia" which had started in the 1880s as a movement to keep Asians out of the continent was codified in the first legislation to be passed by the new federal parliament in 1901. The Immigration Restriction Act required that any new immigrant to Australia pass a dictation test in any language the immigration official chose: Gaelic, for example.[187] This was actually a compromise solution; the Colonial Office in London had objected to the original restrictions specifying that Japanese were not allowed into the country.[188] Such language was embarrassing to London, which was trying to attract Japan as an ally. Instead, colonial officials suggested adopting the dictation provision that had proved so effective in ensuring racial segregation in South Africa. Reinforcing imperial ties by reinforcing Australia's whiteness was perfectly acceptable to British officials; it simply needed to be done more subtly.[189]

As Japanese power in the Pacific grew, so too did Australian antagonism. Threatened both militarily and—they believed—racially, "White Australia" became a rallying cry for politicians and nationalists. The founding conference of the Federal Labor Party adopted White Australia as a primary objective, stating in 1905 that it was committed to "the cultivation of an Australian sentiment based upon the maintenance of racial purity and the development in Australia of an enlightened and self-reliant community."[190] Japan defeating Russia in 1905 in the Russo-Japanese War increased Australian fear, and racial concerns grew for the next thirty years.[191]

War

Imperial wars played a vital role in the development of Australian national identity and to this day hold an important part in national celebrations and memory. At the time of federation in 1901, Australians were fighting for the British in the Boer War. Though Australia's founding moment was one of constitutional deliberation and political participation and was achieved without bloodshed, the Boer War immediately took its place as the initial moment of blood sacrifice and national founding.[192]

The Boer War served as a substitute for a revolutionary moment but was not a deep wellspring of national feeling.[193] The number of participants was small, and the British—fearing with good reason that the Australians were not well trained—had spread them throughout the ranks, hoping to dilute any negative effect they might have on performance. Though there were moments of which the Australian nation could be proud, there were at least as many moments it preferred to forget; discipline was weak and Australians participated in some infamous attacks on civilians.[194] Overall Australians looked to the Boer War as a founding story, but it had little impact on the development of nationalism.

Australian participation in World War I, however, was a defining moment in Australian national identity. Though soldiers fought as members of the Australian *Imperial* Force (AIF), went into battle thousands of miles from home, and had little concern over the safety of their homeland, the soldiers of WWI, and especially those who fought in Gallipoli, came to be viewed as the embodiment of the Australian nation.

Having learned from experience in the Boer War, the Australian government insisted that the soldiers they provided to Britain fight together as Australians, though as noted below they would be commanded primarily by British officers and function as part of the British Army.[195] Thus WWI was the first time Australia fielded a national force in battle.[196]

The soldiers completed training in Egypt, were put under British commanders, and then sent to their first major battle—Gallipoli. Although Gallipoli stands out in Australian military and national history, the AIF actually suffered greater losses on the Western Front. Beginning in April 1916, Australians began participating in the brutal trench warfare in France. At Passchendaele in 1917, Australian forces suffered 55,000 casualties.[197] In the end, Australia sent 330,000 soldiers to fight with Britain abroad: 60,000 were killed and 166,000 wounded or affected by gas.[198] As did the British soldiers they fought alongside, Australian soldiers would take home with them not only their injuries but a belief in the incompetence of the British generals who commanded them in battle.

In addition to fighting for the empire, the fight in Gallipoli validated Australian nationhood, both to themselves and to the other members of the British Empire.[199] British reporters promoted the image of the tough Australian fighter both in England and throughout the empire, and the reputation was quickly accepted.[200] Though serving the empire, Australian soldiers fought together with other Australians. Common memories of home as well as comparison to other imperial troops emphasized the commonality of the Australians and their distinctiveness from their British cousins. World War I thus confirmed Australia's place as a distinct nation in the world, even while it reinforced ties to the empire by shedding Australian blood in imperial battles far from home.[201]

Small-Unit Cohesion

The Australian Army was founded with the Defence Act of 1903, which provided for the establishment of a small army divided into Permanent Forces and Citizen Forces, volunteer forces, and reserve forces.[202] None of these forces could be required to serve overseas unless they volunteered, which remained the case through World War II.[203] Only white Australians were recruited and allowed to join.[204] The army recruited and trained on a mass basis and was not organized to emphasize small-unit cohesion.

When war broke out in Europe in 1914, Australia immediately offered to send a division of 20,000 infantry and a light horse brigade to Britain's aid. However, due to the Defence Act's limitations of the use of Australia's army to within Australia's borders, the proffered division had to be recruited from volunteers. This force was called the Australian Imperial Force (AIF).

The first contingent of the AIF sailed for Suez within six weeks. The Australian government insisted that its soldiers fight together as a national force but did not provide any support or administrative services—only fighting men. As one historian of Australia's army has pointed out, "The AIF was not designed, nor was it able, to fight on its own. Rather, it had to be subsumed into the organization of another army, which would provide for the essential support services."[205]

After World War I, the AIF was quickly demobilized and the army drastically cut.[206] Lack of funding and a belief in the Singapore strategy shaped the organization of the Australian military during the interwar years.[207] Australia's government continuously cut funding for the army from 1920 until 1933.[208] In 1922 funding was so scarce that the universal training system was halted; no longer would all white males between eighteen and twenty-six be trained for even the limited two-week period called for under the Defence Act of 1910.[209] By this time the professional army officers in charge of the army called it a "nuclear structure" only, maintaining the structure of the army formally but unable to use it practically.[210]

Because of slow progress on the construction of the Singapore base and continued reductions in the size of the Royal Navy, in 1933 Australia began increasing the army budget. In 1936 Japan renounced the Washington Naval Treaty and began the speedy expansion of its navy, greatly increasing Australian fears. By 1938 tensions in Europe and fears of Japan had so escalated that Australia reinstituted universal training and began to increase the size of the militia. By June 1939 the militia had 77,000 members and army units were back to regular strength on paper, though they still only trained for two weeks per year.[211]

As in World War I, when Australia declared war on Germany it could not legally send any of its current soldiers to fight. So Australia formed the second Australian Imperial Force (AIF) and called for volunteers. Recruitment followed the pattern of 1914, with divisions being national but each state recruiting contingents locally. By March 1940 over 100,000 men had volunteered to serve abroad despite fears of Japanese invasion.[212] Because Australia was in essence raising two armies—the militia that would remain at home and the AIF that would go abroad—there were significant restrictions on who could volunteer for the AIF.[213] Only some of those who were eventually accepted had gone through the two weeks of militia training required of all Australian men under universal conscription.[214] The first Australian Division sent abroad went with extremely limited training on the expectation that it would complete that training in Egypt under British direction.[215] Later divisions underwent more extensive training before being deployed abroad, but none were combat ready on arrival; both in North Africa and Malaya Australian soldiers required further training in British weapons and tactics when they arrived on scene.

This lack of training meant that soldiers had little time to develop strong small-unit cohesion, which is thought to develop in part through overcoming shared adversity such as intense military training.[216] The division between militia and the AIF meant that there was little esprit de corps in the new divisions that were newly formed and without any history of great deeds through which to encourage unit loyalty. There was also a great deal of movement between units as militia soldiers volunteered for service in the AIF and the AIF divisions were formed. This was not a system designed to develop small-unit cohesion.

Canberra again insisted that the Australian forces were a distinct national force that must fight as a corps and answer to Australia. This time the Australians had experienced officers of their own to put in command of their divisions, and they were able to provide greater support services than in World War I. However, in the field—especially in the first two years in the North Africa—the AIF was deployed as an integral unit of the British Army.[217] Though occasional fits of consultation with Canberra took place, these rarely provided real information or choices. Until the end of 1942 Australian forces fought alongside British forces and were in effect subject to British area commanders.

Democracy

Australia became a self-governing, single entity with the adoption of the Commonwealth Constitution on January 1, 1901. Enfranchising all white citizens, Polity V scores Australia as a 10 from 1901 all the way through World War II.[218] While Australia continued to maintain a close political and legal relationship with Great Britain, it was a self-governing democracy at the time the war began. At the end of 1939 Australia had significant experience in competitive and open executive recruitment for office, constraints on the chief executive, and competitive political participation. In 1911 around 90 percent of Australians were considered literate.[219]

CONCLUSION

This chapter has surveyed the national identity, military organization, and political institutions of the three national groups of interest in this project. British identity was informed by Britain's island geography and country landscape, its democratic and Protestant cultural heritage as well as a belief that it was its duty to spread that heritage around the world through empire, and the idea that Britain often had to save Europe from tyranny. India's national identity was divided among secularists focused on Indian civilization and socialism, Hindu nationalists who drew on the myths and symbols of Hinduism for their identity, and Muslim nationalists who identified with the Muslim Ummah and sought protection of their religious practices and identity. Finally, Australian national identity focused on its own unique geography—harsh and unyielding, and far from "white civilization"—as well as the democratic and religious traditions inherited from Great Britain. Australians were also committed to their place in the empire and helping to spread democracy and "civilization" to the non-white world. Each of these national identities had a different relationship with the goals of the war against Germany and the war against Japan, as I will demonstrate in the next three chapters.

While Britain and Australia were democracies at the beginning of World War II, India was not. Britain and India each had militaries organized to encourage small-unit cohesion. Each recruited soldiers directly into specific units, conducted training and deployment by unit, and worked to develop a sense of esprit de corps unique to each unit. The Australian military raised the AIF from scratch and did not follow the unit recruitment, training, and deployment systems of its British cousins.

NOTES

1. Reiter and Stam, *Democracy at War*, 77.
2. Doherty, *Irish Men and Women*, chap. 1.
3. Kumar, *Making of English National Identity*, 145; Colley, *Britons*, 6.
4. Lowenthal, "The Island Garden," 137.
5. As quoted in Lowenthal, 139.
6. Kumar, 217.
7. Grainger, *Patriotisms*, 271.
8. Lunn and Day, "Britain as Island," 124; Black, *English Nationalism*, 128; Kumar, 211.
9. Lowenthal, 140.
10. Lowenthal, 139.
11. Kumar, 167.
12. Black, 124.

13. Colley, 14.
14. Powell, *Nationhood and Identity*, 12; Grainger, 53; Kumar, 145.
15. Vosler, "Making His Way to the Heart of India," 64.
16. Colley, 22.
17. Colley, 29.
18. Kumar, 201.
19. Powell, 65.
20. Powell, 128.
21. Powell, 8, 30.
22. Kumar, 159.
23. Grainger, 57.
24. Kumar, 179.
25. Powell, 106.
26. Powell, 104.
27. Powell, 106; Grainger, 49.
28. Powell, 96.
29. Rose, *Which People's War?* 239.
30. Kumar, 188–89.
31. Grainger, 49.
32. Kumar, 172.
33. Colley, 130 (italic in original).
34. Powell, 116.
35. Grainger, 125.
36. Kumar, 195.
37. Kumar, 213.
38. Grainger, 106.
39. Grainger, 106.
40. Grainger, 106.
41. Grainger, 147.
42. Vosler, 61.
43. Rose, 244.
44. Rose, 279.
45. Rose, 283. See also Owen, "The Cripps Mission of 1942," 75.
46. McDonnell, *Masters of Empire*, 161.
47. Colley, 5.
48. Powell, 117.
49. Powell, 123.
50. Kumar, 198.
51. Grainger, 307.
52. As quoted in Powell, 138.
53. As quoted in Grainger, 323.
54. Kumar, 232.
55. Fussell, *The Great War and Modern Memory*, 343.
56. French, *Military Identities*, 47.
57. French, *Military Identities*, 44.
58. French, *Military Identities*, 46.
59. French, *Military Identities*, 46.

NATIONAL IDENTITY, DEMOCRACY, AND COHESION   63

60. French, *Military Identities*, 63.
61. French, *Military Identities*, 64.
62. French, *Churchill's Army*, 22; Bond, *British Military Policy*, 59–60.
63. French, *Military Identities*, 94.
64. French, *Military Identities*, 83.
65. French, *Military Identities*, 79. See also Murray, "British Military Effectiveness in the Second World War," 128.
66. French, *Military Identities*, 62–63.
67. French, *Military Identities*, 108.
68. French, *Churchill's Army*, 125; Bond, 62; Shils and Janowitz, "Cohesion and Disintegration," 297.
69. Bond, 58.
70. French, *Churchill's Army*, 145.
71. Polity Project, *Polity5: Political Regime Characteristics and Transitions, 1800–2018*.
72. Lloyd, "Education, Literacy and the Reading Public."
73. Chaterjee, *The Nation and Its Fragments*, 6.
74. Raghuvanshi, *Indian Nationalist Movement*, 67.
75. Raghuvanshi, 101.
76. Maclean, *A Revolutionary History*, 145.
77. Maclean, 145.
78. Seal, *The Emergence of Indian Nationalism*, 280.
79. Seal, 327.
80. Seal, 300.
81. Sarkar, *Modern India*, 177.
82. Chaterjee, 12.
83. Gandhi, *Hind Swaraj*, 14.
84. With the exception of Subhas Bose, who is discussed on more detail in chapter 4.
85. Sarkar, 371.
86. Sarkar, 371.
87. Maclean, 104.
88. Kaura, *Muslims and Indian Nationalism*, 13.
89. Gandhi, 31; Raghuvanshi, 64.
90. Maclean, 104.
91. Sarkar, 252 and 356. See also Maclean, 111.
92. Kaura, 56.
93. Quoted in Kaura, 110.
94. Gandhi, 31.
95. Minault, *The Khilafat Movement*, 102.
96. Metcalf and Metcalf, *Concise History*, 174.
97. Gandhi, 37.
98. Gandhi, 14.
99. Gandhi, 35.
100. Gandhi, 37.
101. Gandhi, 65.
102. Gandhi, 32.
103. Raghuvanshi, 6.
104. Cohen, *The Indian Army*, 73.

105. Perry, *The Commonwealth Armies*, 116.
106. Hasan (ed.), *Toward Freedom*, 268.
107. Gould, *Hindu Nationalism*, 102.
108. Imy, *Faithful Fighters*, 5.
109. Gould, 83.
110. Gould, 102.
111. Gould, 90; Sarkar, 298 and 328.
112. Gould, 89.
113. Gould, 48.
114. Gould, 58.
115. Gould, 64.
116. Sarkar, 356; Metcalf, 189.
117. Gould, 59.
118. Gould, 55.
119. Maclean, 103 and 120.
120. Maclean, 120.
121. Gandhi, v.
122. Gould, 5.
123. Gould, 38.
124. Kaura, 4.
125. Metcalf, 203.
126. Kaura, 22–23; Minault, 70.
127. Kuara, 22–23; Minault, 70.
128. Kuara, 18.
129. Edwardes, *British India*, 190; Metcalf, 156.
130. Kuara, 19.
131. Kuara, 23.
132. Minault, 129.
133. Minault, 148.
134. Minault, 169.
135. Minault, 169.
136. Kaura, 142.
137. Raghuvanshi, 62.
138. Raghuvanshi, 137; Imy, 56–57.
139. Kaura, 21.
140. Kuara, 36.
141. Kuara, 19; Raghuvanshi, 110.
142. Hasan, 318.
143. Imy, 22–23.
144. Imy, 23.
145. Cohen, 30.
146. Perry, 82.
147. Cohen, 59; Imy, 5.
148. Cohen, 44.
149. Cohen, 42.
150. Roy, *Battle for Malaya*, 32.
151. Cohen, 55.

152. Perry, 98.
153. Cohen, 53.
154. Imy, 143
155. Cohen, 44.
156. Cohen, 43.
157. Barua, *Gentlemen of the Raj*, 15.
158. Barua, 16.
159. Cohen, 83; Barua, 52.
160. Cohen, 121 and 123; Barua, 56–57.
161. Barua, 81.
162. Imy, 156.
163. Perry, 115.
164. Barua, 103.
165. Edwardes, 200, 157.
166. Sarkar, 336–38,
167. Chaudhary, "Caste, Religion and Fragmented Societies."
168. Macintyre, *A Concise History of Australia*, 130.
169. Macintyre, 131.
170. Macintyre, 133; McMinn, *Nationalism and Federalism*, 96; McLean, *The Prickly Pair*, 95.
171. Younger, *Australia and the Australian*, 585.
172. Younger, 432; McLean, 64.
173. As quoted in Hirst, "Empire, State, and Nation," 141.
174. Younger, 433.
175. McMinn, 7.
176. McMinn, 20.
177. Macintyre, 147.
178. Macintyre, 165.
179. Hirst, 160.
180. McLachlan, *Waiting for the Revolution*, 228.
181. Eddy and Schreuder, *The Rise of Colonial Nationalism*, 135.
182. McMinn, 116.
183. Younger, 585 and 587.
184. Younger, 494; Macintyre, 170.
185. Hirst, 160.
186. McMinn, 122.
187. Welsh, *Australia: A New History*, 341–42.
188. Welsh, 342.
189. Macintyre, 142.
190. Macintyre, 143.
191. Younger, 440.
192. Eddy and Schreuder, 153.
193. McKernan and Browne (eds.), *Australia: Two Centuries of War and Peace*, 157.
194. McKernan and Browne, 155.
195. Mckernan and Browne, 158.
196. McMinn, 226–27.
197. Welsh, 376.
198. Younger, 468.

199. McKernan and Browne, 161.
200. McMinn, 226.
201. McMinn, 227.
202. Palazzo, *The Australian Army*, 26.
203. Palazzo, 26.
204. Johnston, "The Civilians Who Joined Up, 1939–1945."
205. Palazzo, 67.
206. McKernan and Browne, 224.
207. For further discussion of the Singapore strategy, see chapter 4 (Malaya).
208. Palazzo, 101.
209. Palazzo, 106.
210. Palazzo, 107.
211. Palazzo, 128.
212. Palazzo, 146.
213. Grey, *A Military History of Australia*, 146.
214. Grey, 147.
215. Grey, 155.
216. Ben-Shalom et al., "Cohesion During Military Operations."
217. McLean, 116.
218. Polity Project, *Polity5*.
219. Burke and Spaull, "Australian Schools."

CHAPTER 3

North Africa

This chapter begins with a brief overview of the political and military situation in North Africa, primarily the area from Cairo, Egypt, to Tripoli, Libya, from 1940 when the Germans first enter Libya to late 1942 when the Australians are sent back to the Pacific. It considers issues of equipment and training, which may affect levels of will. It then examines each national group in turn, exploring the perceptions of compatibility between the war goals and a group's national identity, and then looks at the group's level of will in combat throughout the eighteen-month period. Each section also considers the national group's small-unit cohesion, democratic practices, and level of perceived threat. This chapter concludes that all three national groups had high levels of will to fight against the Axis forces, demonstrating high morale, discipline, and initiative. However, some Indian units threatened discipline by informing their officers that they would refuse any order to carry steel helmets as a violation of their religion. Indian will was thus threatened at some points over ethno-religious identity. National identity theory explains both British and Australian will to fight. British and Australian national identity led those soldiers to perceive the war against Germany as one to defend their way of life—including democracy—against a totalitarian threat. In that way national identity is a necessary precondition for understanding threat at all. Secondary cohesion best explains Indian will, but it appears that when not coupled with national identity, small-unit cohesion is a brittle basis for will. It can turn into a force against discipline should the unit's identity come into conflict with the military's larger goals or strategy. These factors thus inform one another and are interrelated.

BACKGROUND

When Britain declared war on Germany in 1939, it was necessary to reinforce the British garrisons in North Africa that would protect oil supplies from Axis attack. Though Britain had minimal territorial holdings in the region—the mandates in Palestine and Jordan were its only official colonies—it had military agreements with both Egypt and Iraq in order to protect oil supplies. North Africa was the focus of the Western Allied military struggle against Hitler until 1943.[1] There were British forces stationed in the region, and contingents from India, Australia, New Zealand, and South Africa joined them. More British units were sent to the region after evacuating Europe at Dunkirk.

When Italy joined the war on Germany's side on June 10, 1940, it proceeded to attack British forces in Somaliland and the Egyptian frontier with Libya.[2] After the withdrawal from those areas and consolidation, the British counteroffensives throughout Africa and North Africa in Autumn 1940 met with huge success, driving Italian forces out of the areas Italy held prior to the war.[3] By December 1940 the Desert Army (as British Imperial forces were known at the time) controlled the area from Egypt to Tripoli and were poised to push the Italians out of North Africa entirely.

At this point two political leaders intervened in the military situation. Hitler agreed to Mussolini's request for help and sent German panzer divisions under command of General Erwin Rommel to Tripoli to retake the Italian holdings. Hitler also agreed to aid the Italians in their ill-fated invasion of Greece, leading Churchill to insist on upholding the promise to send troops to aid Greece against German invasion.[4] Therefore, in late March a much depleted British force—having sent a large number of its troops and equipment to Greece (see chapter 5)—faced not only the remainder of the Italian forces in Libya but also fresh German units led by Rommel.

The story of the North African Campaign is one of gains and losses by each side as armies pushed each other back and forth over the desert. Supply lines were not secure for either side, though Axis forces held the advantage of controlling the Mediterranean for much of the period. Both armies were frustrated by their leaders' distraction in other regions—the British in Greece and later the Pacific, the Germans in the Soviet Union.

Imperial forces also underwent considerable changes during the course of the campaign, in national makeup, organization, and leadership. General Wavell led the Desert Army from 1939 until July 1941 when Churchill replaced him with General Claude Auchinleck in an effort to spur more aggressive action. Auchinleck proceeded to reorganize his forces into the Eighth Army, which would fight the Germans across North Africa and into Italy over the next three years. Auchinleck was replaced by Bernard Montgomery in August 1942. British, Indian, and Australian units were present in North Africa from 1940 until the end of 1942 when the 9th Australian Infantry Division returned to Australia to participate in the fight against Japan.

Imperial forces were consistently outgunned by German equipment throughout this campaign. What armor and artillery there was belonged to the British armored units, though there were a few motorized Indian and Australian units as well. However, the larger problem lay in generalship and doctrine that failed to make the best use of the armor and forces that were available.[5]

All troops received most of their training in North Africa before being sent to the desert lines.[6] All units received basic training in their country of origin but expected to be trained for their specific assignments once they arrived in theater. When they embarked, the troops had no official word on what their final destination would be; it could be North Africa, Hong Kong, or Burma. Once they arrived, they would generally receive several weeks' training in the tactics and doctrine of the region where they were to fight.[7] Thus all the imperial forces in North Africa—British, Indian, and Australian—had the same tactical training in the same military doctrine.

BRITISH IN COMBAT

Perceptions of the War Goals

The British people initially supported Chamberlain and his efforts to achieve a peaceful settlement with Hitler during the 1930s. As the British government slowly came to the conclusion that it must take a stand against German expansion in Europe, Britain also began to accept the necessity of another war with Germany. By the time the Desert Army engaged Rommel in North Africa, the British people had suffered through months of intensive bombing and a period of fear of imminent invasion.

According to scholars of the Second World War, the British people saw several main issues in the fight against Germany and Italy:

> First, acute fear of the nation's enemies; second, mass mobilization, which revived the economies of Scotland, Wales, and the north of England; third, unprecedented solidarity between classes and sexes, which led to the creation of social democracy; and fourth, the factor least discussed by historians, national guilt about appeasement.[8]

The British people were directly threatened by German aggression, though the threat of invasion receded after May 1941. The practicalities of the war drew them together despite class lines that remained after the First World War. But in addition to fear and solidarity, the British recognized the failure of appeasement and the tyranny that had spread over Europe as a result. Churchill made repeated reference to "honor" and following through on commitments to allies in his speeches, and was sure to note when others failed to do so.[9]

In addition to themes of honor, Churchill also referenced Britain's past wars in defense of the homeland and against tyranny in Europe. After Hitler invaded the Soviet Union, Churchill offered an analogy with Napoleon, emphasizing Britain's repeated role as savior of Europe:

> Napoleon in his glory and his genius spread his Empire far and wide. There was a time when only the snows of Russia and the white cliffs of Dover with their guardian fleets stood between him and the dominion of the world. Napoleon's armies had a theme: They carried with them the surges of the French Revolution. Liberty, Equality, and Fraternity—that was the cry.... But Hitler, Hitler has no theme, naught but mania, appetite and exploitation.[10]

Churchill thus evoked British myths of national identity as a tolerant democracy and being the last best defense of Europe against tyranny as the lens through which to view the war against Hitler.[11]

The Labour Party also saw Hitler as a threat to a mythical British way of life—a way of life based on freedom and self-government for all peoples (including the British colonies). According to *Why We Fight: Labour's Case*,

> We believe in liberty, through which alone the mind and soul of the peoples of the world can find free expression. All peoples, whether they be great powers or small nations, have a right to live in security and independence, without threats or menaces or the use of force ...

we assert the right of all nations to live their own lives. We deny the right of any power to commit acts of brigandage or to seek to attain its ends by means of force or the threat of force.[12]

The book goes on to point out that Hitler had made a mockery of democracy in Germany, started a war despite Britain's best efforts, and that "We are fighting ... for a quite simple principle—that of liberty."[13] It was not only Churchill and his grand oratory that portrayed the war against Germany as one to defend freedom and democracy across the world. Labor, part of the national government but often opposed to Churchill, also affirmed those principles as the high goals of the war. Labor's rhetoric focused more on the "march of progress" that Hitler was attempting to halt, but both conservative and progressive politicians used the symbols of British history and democracy to frame the goals of the war against Germany.

The government's message that this was a war to defend Britain as well as British principles resonated with the people. Gallup polling for the period indicates that the British population was committed to the war and to its goals. In June 1941 when asked "what do you think we are fighting for?," 46 percent answered "freedom, liberty, and democracy," while the next-largest category was "to stop Fascism, Hitlerism, Nazism, aggression" at 14 percent. Of those who gave these reasons, 73 percent and 69 percent respectively said they believed Britain was going about the war in the right way.[14]

Only 5 percent of Britons surveyed answered "for our existence, our lives," despite the fact that in that same poll, 39 percent said that they thought that Germany would try to invade Britain that year.[15] When asked if they would approve or disapprove of peace talks with Germany at that time, 82 percent said they would disapprove.[16] Threat played a role in British commitment to the war but was understood through identity.

Not all soldiers described their motivation in terms of identity. In explaining his motivation in a letter home, one soldier in North Africa wrote of the bonds of comradeship and small-unit cohesion: "Already we begin to fight. Not because we want to liberate Europe from its Nazi bonds; not because we think our ideology is better than theirs; but because we have seen our comrades dying by our sides, because our relatives have been blitzed at home; and we fight to kill, to exterminate those wicked men from the earth."[17] Fear, anger, and comradery did motivate soldiers in battle. Yet even after the direct threat of invasion receded and the worst of the Blitz was over, the British continued to believe that they must turn Hitler back at any point they could. Except for a brief campaign in Greece, the main fight with Germany took place in North Africa from March 1941 until the German Army evacuated to

Sicily in 1943. Japan's entry into the war does not seem to have made a strong impact on British perceptions of the war with Germany, except to reinforce the idea that this was a world war for freedom and democracy.[18] For the British, North Africa was still the primary front in that war.

Race also played a role. Soldiers stationed in Palestine wrote home of their frustration with the high prices charged by both Arab and Jewish residents. Antisemitic comments were not uncommon, with one soldier going so far as to write "Hitler is quite right: there is only one way to treat the bloody Jews."[19] British opposition to Hitler did not negate the racism that made up part of their national identity.

In the field, British troops paid attention to the broader war effort, and Germany's invasion of the Soviet Union increased their sense of determination. According to censorship reports,[20] the week after Germany invaded the USSR: "The chief topic in correspondence from all ranks this week has been the German-Russian war" and "to the men in Tobruk it came as quite a tonic."[21] As the war on the Eastern Front progressed, British soldiers wrote home of the possibility of making a move now, while the Germans were distracted. Additionally, imperial success in Syria increased morale over the course of the summer.[22] British soldiers were also generally aware of current events and their relation to the war. "The developments in the Far East have aroused considerable speculation amongst the members of both Imperial and British forces."[23] While the Australians and New Zealanders were particularly concerned about Japanese proximity to their own homes, the British were more interested in whether it would draw the United States into a more active role, thereby increasing the pressure on Germany.[24] Soldiers paid attention to politics, and it influenced their motivation.

British soldiers were interested in the goals of the war as well as its progress. In August 1941, Churchill and Roosevelt's meeting in the Atlantic and the contents of the Atlantic Charter agreed there were made public. According to one soldier, "I am very optimistic about the future. The 'Eight Points' set out by America and England, *just about reach the standard of ideals that we are fighting for.* It will apply to individuals as well as nations and I think that when these Eight Points become known to the ordinary people of the Axis centres, they will lose their desire to carry on a useless fight, that, if they won, would bring them only misery."[25] Censorship reports indicate that the Atlantic Charter was a topic of comment in letters home for some weeks after it was announced.[26] Soldiers in battle zones paid attention to discussions of the goals of the war, and the principles of self-determination and economic progress set out in the Atlantic Charter resonated with them.[27]

In addition to the political goals of the war, British soldiers referred to memories of England, especially its geography and countryside, to explain why they were fighting: "Derbyshire conjures up a thousand memories, riding with you, walking in those hills and dales. This is what we are fighting for, England the scene of our tours."[28] Home, and the hills and dales believed to be unique to English geography, as well as politics, inspired soldiers to keep up the fight.

The British believed the war with Germany to be compatible with several aspects of their national identity. It was a war against tyranny and oppression, a war to protect the democracy and freedom that had been elements of British identity myths for centuries. They also fought to protect small nations in Europe and the principle of self-determination, highlighting their identity as savior of Europe as well as the contradictions of their national myths. Finally, they fought to protect their own homeland from destruction in the German bombing campaign and potentially even invasion. The British believed there was strong compatibility between the goals of the war and their own national identity. In North Africa the British saw a chance to engage the Germans directly and protect their strategic supply lines and oil supply, setting themselves up for victory.

The Campaign

The first encounter the Desert Army had with its German opponent came at the end of March 1941. Imperial forces had succeeded in driving the Italians back to Tripoli and had settled in to hold the position while many were sent to Greece (see chapter 5). Some reinforcements moved to the front, but what re-equipping took place was a matter of salvaging Italian tanks and guns rather than adding new British equipment.[29] The British forces that remained in North Africa were underequipped and spread thin across a wide front, with very few defensive strongholds and a very long supply line.

One of the first units to engage the German forces was the Tower Hamlets Rifles: "These Cockneys were mostly Territorials—weekend soldiers, called to the colours on the outbreak of war."[30] The Tower Hamlets were from the same general area of East London but were not part of one of the historical regular army units with their traditions and history. They had originally been part of the Territorial Army, essentially a reserve army made up of middle-aged men who trained a few weekends per year. After the Regular Army was expelled from Europe at Dunkirk, the Territorials were called up and charged with either defending the home island from invasion or fighting the Italians and Germans in North Africa.

Despite their lack of intensive training and esprit de corps, the Tower Hamlets demonstrated strong morale and discipline in the first major engagement with the Germans. On March 30 and April 1st, the Germans attacked the positions near Mersa Brega where the Tower Hamlets had been ordered to make a stand: "The Support Group at Marsa Brega were confident. The battery commander with the Tower Hamlets reported to his commanding officer that the infantry were quite happy and intended to stay where they were."[31] Even the dreaded Stuka dive-bombers did not fluster the Tower Hamlets, who responded with machine-gun fire and cries of "Take this for London, you bastards!"[32] This was the Londoners' first chance to engage with the forces that had been bombing their homes for the last eight months, and they took advantage of it.

When the Panzer Grenadiers attacked their position, the Tower Hamlets held off the first wave of the assault but were forced back when German tanks were able to drive a wedge between two of their positions.[33] A unit from the Support Group was able to stage a counterattack that confused the Germans long enough for the Tower Hamlets to extract themselves and withdraw into the town with the rest of the British forces.[34] This Territorial unit demonstrated morale and discipline in the face of a better-equipped enemy and slowed the German advance into Libya and through the Desert Army. The Tower Hamlets saw themselves as actively participating in the defense of their homes, despite being on another continent.

On April 3, the larger Support Group, of which the Tower Hamlets were a part, was forced to retreat from Mersa Brega back toward the rest of the Desert Army. They were ordered to retreat rather than to become heavily committed to battle as General Wavell feared the loss of what little military force remained in North Africa to defend Egypt. Rather than flee and leave the German forces intact to attack Benghazi—the next city in their path—the Support Group chose to institute a "fighting retreat" and sought to destroy as much of the enemy force as possible as they moved eastward.[35] In this way the British forces at Mersa Brega demonstrated both discipline and initiative—following orders to retreat in an orderly fashion but still seeking out engagement with the enemy and inflicting as much damage as possible while obeying those orders.

On April 6 the Desert Army decided to take up a defensive position at the old Italian fortress of Mechili with an assortment of troops. Australian, Indian, and British soldiers all converged on the fortress just before it was surrounded by the German 5th Light Division.[36] Many of these soldiers were from units that had already engaged the Germans elsewhere and been forced to retreat. They brought with them whatever equipment they still had and assembled new groups of soldiers to make use of that equipment and what defensive structures the fortress provided.[37] General Gambier-Perry, commanding one of those units (a unit in name only by the time it arrived at Mechili), took command of Mechili on his arrival.[38] The fortress was out of contact with Desert Army headquarters, and the leadership did not know whether to expect reinforcements.[39]

By April 7 the Germans had surrounded this motley group, and Rommel himself was present. He twice demanded the fortress surrender and was twice refused.[40] All units endured artillery shelling with determination and awaited orders to break out through the surrounding German forces. When the breakout orders were finally given, the resulting chaos was due in great part to poor communication and leadership at the top. According to the official historian, if the plan developed had been communicated and adhered to, most of the forces in Mechili would have escaped to rejoin the Desert Army in the Western Desert. However, poor communication between armor and infantry as well as poor leadership by General Gambier-Perry resulted in the majority of the forces surrendering to the Germans.[41]

Three squadrons—one each of British, Indian, and Australian soldiers—were left behind as a rear guard for the attempted breakthrough. Rather than surrender with the rest of the garrison, they decided to attempt a breakout on their own. These units, led by Major Rajendrasinhji (Indian) and Captain Dorman (British), demonstrated initiative and the discipline to act together despite their multinational makeup.[42] The British unit had been together for some time, but it had not worked with its Indian or Australian counterparts for more than the few days they had been in the fortress together. Yet they worked together, charging the German line at full speed and surprising the German gun crews.[43] The multinational unit swung east and eventually connected with a larger British battalion. As they made their way back, they even captured prisoners along the way.[44] Small-unit cohesion and solid leadership helped these units to demonstrate the discipline and initiative necessary to break through the German lines, though all three were able to work together without having served or trained together before. The remaining 2,000 soldiers at Mechili surrendered to Rommel with General Gambier-Perry.

Most of the British elements of the Desert Army made their retreat eastward after April 8, stopping at the Gazala desert just east of Tobruk, a fortified port city. However, some British forces entered Tobruk with the Australians and participated in its defense for the next few months. Most of the British troops in Tobruk were from artillery or armored units, but had lost their equipment during the retreat. What vehicles there were could only be used with the utmost judiciousness as fuel was scarce.[45] British cavalrymen became infantry and were dispersed throughout the defenses, which consisted of concentric rings of tank ditches, with an outer perimeter made up of field posts: three concrete circular weapons pits sited at ground level and connected by underground passages.[46] Some British squadrons that still had tanks acted as mobile armored squads, attempting to get to the spot where they could be the most useful whenever the Germans attacked. Demonstrating initiative, the British artillerymen taught themselves how to use the old Italian guns captured when the Italians abandoned the fortress in December 1940; they were nicknamed the "Bush artillery" for their self-taught tactics and homemade gun arrangements.[47]

When the Germans attacked Tobruk, they expected their tanks to have the same effects on imperial forces that they had on European forces during the battles in France. Instead, the infantry refused to be startled by them, and the British artillery and what armor there was met the attacks and stopped them.[48] That British artillery was assembled in the field; it bore little resemblance to the original organization of the British armor. The men had not been trained on most of the equipment they used, and they were dispersed among Australian infantry with whom they had not trained or even worked before.[49] Despite the inattention to small-unit cohesion, British soldiers were able to adapt and fight with discipline and initiative in new units doing new jobs.

British leaders encouraged their soldiers to think of their fight and trials in Tobruk as part of the wider war effort. One officer, Captain Graham Slinn, explained to his men that if the port at Tobruk fell, the Germans would be able to conquer Egypt and cut off Britain's oil supplies. Drawing on the myth of British wars to defend the homeland against European aggression, he sent his men into battle with the final words of Shakespeare's *Henry V* speech at Agincourt: "You know your places; go to them and God be with you" and told them the fate of England and the free world was in their hands. The unit was especially urgent in their shelling of German positions during the battle that evening.[50] One British soldier, referencing the press that they were getting, wrote home: "This place will go down in British History. I've been here seven and a half months now and I'm proud of it although I've had quite a lot of narrow escapes. Don't worry."[51] The sense of participating in the creation of British history and national identity was motivating under difficult conditions.

As the initial battles around Tobruk settled down into an active siege, British armored soldiers demonstrated initiative in a number of ways. They took it on themselves to train the Australian infantry in tactics for tank warfare. They taught the Australians to lie down in their trenches and let the tanks drive over them, waiting for the infantry that followed.[52] Some of the British armor, without fuel for their vehicles, even decided to join the Australians in their patrol work: "Thirsting for action, Tank Regiment officer Rea Leakey, now an acting major, attached himself to an Australian infantry battalion as an acting corporal and went out one night with two privates to man a concealed artillery observation post over a mile into no mans land."[53] The three were fired on for some time before returning to their own lines. Other British dismounted artillerymen occupied perimeter posts and actively participated in patrolling no-man's-land, interfering with enemy defensive construction and even causing the Italians to fire on themselves.[54]

Morale in Tobruk was generally good, though being confined to the twenty-eight-mile perimeter was wearing on all present. At the beginning of the siege, the fort's commander, Australian General Leslie Morshead, announced there would be no retreat; they would hold the fortress until the siege was broken.[55] The knowledge that they were committed to Tobruk's defense and that the defense was vital for victory over the Germans helped to keep up morale, though not necessarily physical strength. According to Morshead, "the troops were tiring; their health was good but medical officers had noticed that their stamina and their powers of resistance were weakening; this applied equally to British and Australian troops."[56] Morshead's clarity to the troops that they would not retreat in the face of the initial German attacks had improved everyone's morale after their retreat from Libya, and there was a general sense of determination throughout the garrison, despite difficult physical conditions.[57] On learning that his unit would not be relieved, one soldier wrote home: "He [the officer in charge] was no more sorry than we were, and he didn't know how we felt; had he done he would probably have put us all under arrest for threatening to desert. Still, we took it pretty hard but not enough to make us lose that true British spirit."[58] No one enjoyed defending Tobruk, but soldiers connected the duty to their British identity and carried on.

Though the British and Australian soldiers in Tobruk were able to cooperate well in battle, there was some bad feeling between the two groups over the amount of press the Australians were getting as the "defenders of Tobruk." The censorship reports offer an example of such frustration: "Another sore point with the British soldier is that the colonial troops appear to get most of the publicity, and more references in the press and on the radio to the work done by the 'Tommies' would certainly help to cheer the latter up."[59] This irritation with over-coverage of Australian feats exacerbated an overall frustration with the BBC. As one soldier put it: "Do they think they are broadcasting to morons?"[60] All imperial soldiers complained of the mischaracterizations of their battles on the BBC and preferred the German radio station's entertainment to that of their own national broadcasts.[61] British soldiers recognized propaganda when they heard it and dismissed that which did not line up with their firsthand knowledge. Propaganda disconnected from soldiers' reality was not successful. At least the German station played music.[62]

After two brief and unsuccessful attempts to relieve Tobruk over the summer, Churchill replaced General Wavell with General Claude Auchinleck of the British Indian Army. Between the stalemate battle Battleaxe in June and Auchinleck's first offensive, Crusader, in November, both the imperial and German forces were reinforced and resupplied.[63] At this point the Desert Army was renamed Eighth Army. It was also during this period that censorship reports indicate a discussion of the Atlantic Charter and the war aims laid out therein.[64] British soldiers were thinking about the principles of self-determination and economic progress to which the charter committed and were buoyed by the increased support from the United States. Though Crusader was not the definitive battle that Churchill had long hoped for, it did succeed in relieving Tobruk and pushing the Germans further west.[65]

Soldiers knew that a new push was approaching and were preparing themselves. References to British identity appear in numerous letters quoted in censorship reports. Describing German propaganda efforts in July, one soldier wrote home: "Jerry dropped leaflets on us a few days ago calling on us to surrender and wave white flags etc. What did we do? We laughed like hell at the joke and the compliment. . . . It shows that he cannot understand the determination of an Englishman's mentality or appreciate his sense of honour."[66] At the beginning of October one wrote home, "We'll have to fight but when the next big push does come it will be the beginning of the end . . . we realise only too well what we are fighting for. . . . It's fighting to preserve England—our homes—wife and family, etc." Another wrote, "I shall be glad when the war is over, so as I can come home to you but I wouldn't have missed it as I know what we are fighting for is right and should there ever be another war I should go just the same, as I have always said if a country is worth living in it's worth fighting for and believe me England's worth everything."[67] Yet another soldier referenced the Tommys of WWI, noting "these lads are as good as their dads were and no mistake."[68] The motivation of fighting for hearth and home, and the hope that this might be the real breakthrough, helped sustain morale.

Crusader began on November 18, 1941 with the newly organized Eighth Army headed toward the German forces with very high morale.[69] Those who had been in North Africa for some time were ready to exchange boredom and training exercises for substantive action, and those who were new to the war were ready to take their crack at the Germans.[70] In examining letters sent home the week before Crusader began, censors wrote: "Expectancy, cheerfulness, confidence and determination were the outstanding characteristics of this correspondence."[71] A large number of British soldiers told their correspondents not to worry if they did not hear from them for some time as they "would have less opportunities for letter writing than in the past," irritating censors by hinting at the expected action.[72]

Some of the troops that participated in Crusader, such as the 22nd Armoured Brigade, had arrived as late as October, barely having time to refit their tanks for the desert conditions before going into battle.[73] Again, these troops had been Territorials, and their training up until October 1941 had concentrated on defending Britain from German invasion—a very different job than fighting a desert war.[74] Nevertheless, the 22nd fought with morale and discipline in their first desert actions. Despite a quick defeat in which they lost forty of their tanks—and discovered their disturbing tendency to catch fire—the 22nd regrouped quickly and went back into action.[75] The newly arrived 22nd, despite lack of training and experience with their new tanks, continued to engage the German panzers until they withdrew.[76]

More experienced armored units, such as the 7th Armored Support Group, were more successful in doing damage to the German armor but continued to suffer from their general's tendency to separate them, thus sapping their strength.[77] Strong leadership in the field helped, and the 7th Support Group proved itself capable of adopting new tactics quickly given direction. Brigadier Jock Campbell, seeing that destruction of his Support Group was imminent should they continue to engage the enemy piecemeal, rallied his tanks. He stood up in the back of his staff car, forming the remaining tanks into what came to be known as "Jock Columns," and led them back into the battle.[78] Brigadier Campbell was able to devise a new tactic based on his observation of German tactics, and the men of the 7th Armored Support Group were eager to try it out on their opponents. The British thus demonstrated morale and initiative in battle, ready to continue the fight even after devastating losses and to try new tactics against the enemy.[79]

British infantry were more active in the breakout from Tobruk than in the battles in the open desert. The 70th Division, which had replaced most of the 9th Australian Division in Tobruk in August and September 1941, was tasked with attempting to break through the German and Italian siege to meet the imperial forces attacking from the east.[80] Again, communication was poor and the generals issuing orders—including the order for the 70th to attempt a breakout—did not have full information regarding the entire battle.⁸¹ However, the 70th successfully broke through the first German line around Tobruk and established a salient in which they hoped to meet their fellow imperial forces who were to break through from the other side.⁸²

British esprit de corps exercised influence in this case, when The Black Watch, a nominally Scottish unit with a history of some 200 years, attacked a position they believed the Italians occupied. The German soldiers actually defending the position put up a staunch fight. The men of The Black Watch were rallied by one of the company officers, who stood up despite the incoming fire and shouted "Isn't this The Black Watch? Then come on!"⁸³ The Black Watch, among other British units, succeeded in taking the position. According to the official Australian history, "The action developed into a muddle redeemed by great leadership and utmost bravery."⁸⁴ British esprit de corps and leadership rallied the troops in the face of a strong opponent and helped to gain an important position.

The British continued to hold this salient for some days longer than they had originally expected and with very little armored support against a well-armored enemy, waiting to meet up with their relief. Tobruk was finally relieved when on November 26, 1941, a small group of New Zealanders linked up with the 70th Division's salient and helped them to establish a corridor to the Eighth Army. Though Rommel attacked this corridor the first week of December, hoping to again cut Tobruk off from the main British force, the 70th held the position. The Germans began their westward retreat shortly thereafter.

In the period after this battle, One British soldier wrote home:

> England was proud of us, and by God, Ted, I was proud of England. All the patriotism that is in me welled up in my throat and choked me. There in the middle of the vast desert it came to me in a flash, this England of ours will live, Ted, when you and I are gone and forgotten . . . live by the blood of its sons. The freedom, kindness, and all that our England stands for is ours which no man can take away.⁸⁵

An emotional statement after a hard-fought victory, one that demonstrates that British soldiers identified with their nation and the myths that defined it.

Rommel, never willing to let a defeat stand, regrouped and attacked the Eighth Army lines in the Libyan Desert in January 1942 while the British were distracted by the Japanese attacks in the Pacific. He succeeded in driving the Eighth Army back to the Gazala line, away from his own ports and supply centers. Before Rommel attacked, censorship reports indicate morale was high. "Despite appalling weather conditions, alternating between dry sunny days, terrific thunderstorms, violent sandstorms and bitterly cold nights, the eternal bully beef and biscuits, shortage of water and mail difficulties, the morale of the forward troops in Cyrenaica appeared to be exceedingly satisfactory."[86] However, after the withdrawal back to Gazala, giving up much of what the Eighth Army had gained in Crusader, the tone of the letters did change. According to the censorship report for 11–17 February: "Considerable disappointment has been expressed by all ranks over our withdrawal in Cyrenaica, and the tone of the correspondence examined in the period under review was not as satisfactory as it had previously been. There was, however, no concern that the Germans would progress further or doubt in our ability to regain the lost ground."[87] Losing all they had so recently won was disheartening and frustrating, but apparently not enough to change soldiers' beliefs about their ultimate victory.

Fennell argues that low morale was a fundamental problem for the Eighth Army during 1942 and explains some of their reverses.[88] However, before the battles began morale was "exceedingly satisfactory." Soldiers went into battle with high morale and were disappointed and frustrated with their losses afterward. The censorship report notes, "While no doubts are entertained as to the final outcome of the Libyan Campaign, there is no denying the fact that the tone of correspondence from British troops in the forward areas was lower than it has ever been in the last three months, and reflected the general disappointment of all ranks over our hurried withdrawal."[89] Soldiers were "browned off" at the idea of yet another summer in the desert—an unsurprising attitude.[90] Yet defeatism does not appear to have been a problem even after a major withdrawal.

In March 1942, one soldier cited small-unit cohesion as a source of confidence, writing: "Well, in two days time a great day will dawn for this regiment. We'll be three hundred years old. . . . I am darned glad I joined it because it's proved itself one of the best fighting units in the Middle East and on top of that it's the best crowd of blokes I could ever wish to meet."[91] Just a week earlier another soldier wrote, "We have lost some good lads but it makes others all the more eager to avenge them. You have often heard that the British soldier is the finest in the world and I assure you their fathers would be proud of them, if only they could have seen them as I have, with tanks . . . we can beat Jerry anywhere."[92] Even after substantial losses British soldiers drew on a combination of small-unit cohesion and national pride as motivation to continue the fight.

Following the retreat to Gazala, both armies settled in to reinforce and resupply. The British established a series of "boxes" along a line from Gazala in the north to Bir Hacheim in the south. Each box was garrisoned with a brigade, which Auchinleck intended to function as an integrated force similar to the Germans—armor, artillery, and infantry all working together. Auchinleck planned to make the next move, and the three months between their retreat to Gazala in February and Rommel's attack at the end of May were spent adopting this new organization and training for the planned offensive. But Rommel acted before Auchinleck could, and once again the British positions were too far apart to be able to support one another in battle.[93] The result would be nicknamed the "Gazala Gallup" by the very troops who participated in the retreat to the prepared El Alamein lines for the last stand before Rommel could enter Egypt.[94]

Once again, though British soldiers did not relish their encounters with the enemy, they were ready to do their job and go home. Of letters written the week before that June battle, the censorship report notes: "The mail contained ample evidence of the troops' fighting spirit, and those men who were indiscreet enough to state that they were expecting something to happen soon, went on to say, 'the sooner the better.'"[95] Additionally, letters indicated that the troops had listened to Churchill's May 10th speech, which described the current state of affairs. Drawing on Britain's myths of defending Europe from tyranny, the speech emphasized that while Britain had stood alone "as has happened before in our island history, by remaining steadfast and unyielding—stubborn, if you will—against a Continental tyrant, we reached the moment when that tyrant made a fatal blunder."[96] Hitler, like the tyrant Napoleon before him, had invaded Russia. "But they, like us, were resolved never to give in. They poured their own blood upon their native soil."[97] Now, with the Americans also joining the effort, Churchill struck an optimistic note that resonated with the army in North Africa. According to the censorship report for May 20–26, "The Prime Minister's speech was the 'high light' of the week and provided much encouragement. 'Aggressively optimistic' was the definition given by an officer, while the following was typical of O.R comment: 'if he can use that sort of tone you can tell things are going well.'"[98] Churchill called on the British myth of standing alone against continental tyranny to frame the difficult fighting behind them and promote an optimistic view of future fighting. That use of national myth resonated with the soldiers.

After a spring of losses, both on their own front and in the Pacific, the troops in North Africa believed that the Russian and American war efforts were finally beginning to turn the tide and there would be no more reversals. They were sorely disappointed by their next action.[99]

Though individual units demonstrated discipline, morale, and initiative, mismatched equipment combined with a poor battle plan led to Eighth

Army's retreat back to the Egyptian frontier by the end of June 1942. When Rommel attacked the Gazala line, the individual boxes found themselves outmatched yet too far apart to receive reinforcement. The first brigade box, 150th, fell at noon on June 1, though most of the brigade was able to break out before it was overrun.[100] This left a large gap between boxes, allowing Rommel to drive a wedge between the defenders and once again address the Eighth Army one brigade at a time.

Yet even one brigade at a time posed problems for Rommel. Several British armored brigades held out after the fall of the 150th Box, continuing to engage the enemy and conduct a fighting retreat. It was only when the Knightsbridge Box, a key feature of the defensive scheme, was surrounded that those brigades were ordered to retreat and the box was abandoned under orders. Despite the nickname of "Gazala Gallop," the British forces fighting at Gazala maintained discipline even when retreating.[101]

As the Eighth Army retreated back toward the Alamein line with each collapsed brigade box, the garrison at Tobruk was once again isolated from the rest of the army. This time the military chiefs had already decided that they would not garrison the fortress for a siege again; with the expanded war in the Pacific, the navy could not undertake the effort to supply an isolated garrison by sea.[102] On June 21 General Klopper, the South African general commanding the fortress, surrendered. A number of soldiers refused to surrender, not having confirmation of the order from Klopper, and sought to escape before the Germans moved in.[103] About 400 British soldiers made it back to the Eighth Army through the Italian lines.[104] Others continued to fight for up to twenty-four hours after the fortress surrendered. The Cameron Highlanders finally emerged from their position on June 22 and marched into captivity with their bagpipes playing the regimental march.[105]

When the Eighth Army regrouped at the El Alamein line in late June 1942, they were "angry, self-mocking and uncertain who to blame for their humiliation."[106] Censorship reports indicate that the dip in morale came through in letters home as well: "The mail did show that the high morale of the troops had suffered a set back, chiefly due to utter physical exhaustion combined with the realization of the horror of battle and the loss of comrades . . . optimism regarding an early finish to the campaign has been tempered by the stiff opposition encountered."[107] Letters compared British leaders poorly to German commanders. "More than one correspondent has stated that 'the 8th Army have more respect for Rommel than for our own high command.'"[108] This goes against what democratic effectiveness theory, which places an emphasis on skilled leadership, would expect. Yet, despite fighting in cobbled-together units with their backs to Cairo and with little confidence in their leaders, expressions of disappointment did not also result in defeatism or despondency about the war. Censorship reports for June 24–30 note,

> Correspondence from British Troops in battle areas written after the fall of Tobruk showed bitter disappointment at the turn of events in the field, but at the same time the writers assured their people at home that despite temporary setbacks we must not doubt ultimate victory. "It's all very disappointing... but we shall fight back" and "don't worry as its just one of those things which work out o.k. in the end," are typical reactions while a third writer feels it to be a temporary reverse and that "the tables will soon be turned again."[109]

British soldiers continued to fight with morale, discipline, and initiative against the Germans, halting their progress toward Egypt.

Fennell points to surrender and desertion rates during this period as evidence that morale was low. He notes that from the beginning of Rommel's offensive at the end of May until late July, Eighth Army lost 1,700 killed, 6,000 wounded, and 57,000 missing and likely to be prisoners of war.[110] General Auchinleck cabled London at this time, requesting that the death penalty for desertion be reinstated. Auchinleck claimed that the rates of desertion and unnecessary surrender had become untenable and that only the deterrent of the death penalty would stem the tide.[111] Indeed, numbers of arrests for desertion, as well as the rate of missing/surrendered as a percent of casualties had increased.[112]

However, historians have pointed to several problems with taking these numbers as indicative of the state of morale or discipline in Eighth Army. First, this was not the first time that a British commander had asked that the death penalty be reinstated since the start of the war.[113] Military commanders had fought the political decision to eliminate the death penalty in 1930 and had sought to reinstate it numerous times since then. Auchinleck's request was one of a pattern of British generals pushing back against a policy imposed by civilians.

Second, when asked to send more precise figures regarding desertions, Auchinleck had to admit that he could not do so until after the fighting was over as, "units had become so intermingled that it was impossible to distinguish between deserters and stragglers who had lost contact with their own unit and attached themselves to another one."[114] It is thus not even clear that rates of missing/surrendered and deserters were actual surrenders or desertions or if these soldiers had simply joined another unit when they became separated from their own, something that happened regularly in the desert.

Finally, uninjured soldiers surrendering in the desert is not necessarily indicative of low morale. As noted above, soldiers often surrendered to the enemy without themselves being injured after being cut off from their supply line and using up all their ammunition.[115] Soldiers were sometimes able to break out of a position and attempt to rejoin the army, but even when uninjured this was not always possible; often those that initially escaped were captured days or even weeks later.[116]

Moreover, POWs regularly escaped back to their lines. Censorship reports for June 1942 highlight information about treatment of prisoners by Germans and Italians based on letters home from those who had escaped from custody. One letter home quoted at length stated, "I have lost four more killed—poor chaps. We gave the Hun a bit of a knock though all was set fair for Cairo—four of my chaps were prisoners and escaped—said the Hun was very decent to them—asked what sort of place Cairo was, they said rotten go to [sic] Alex."[117] A number of letters in this report indicate their authors had been taken prisoner for a short period before making an escape back to the British lines.

The patterns of surrender and escape in North Africa provide support for Grauer's argument that soldiers surrender when they believe they are more likely to survive through surrender and expect their imprisonment to be of short duration.[118] British soldiers tended to surrender when cut off from supply lines and having used up their ammunition. Because they knew from their own comrades that they would be treated decently and their chances of escape were high, they saw surrender as reasonable under the circumstances. Importantly, though, soldiers expected to return to the fight eventually. They do not appear to have connected surrender with quitting the fight as a whole.

British officials were deeply concerned about surrender rates, but they should be understood in the context of desert fighting, other indicators of morale, and the rates at which prisoners escaped and returned to their own lines. The desertions that concerned Auchinleck were not necessarily desertions and were less to do with indiscipline and more to do with the chaos of an army retreating across the desert.

At the end of June, when they arrived at the Alamein line, British units had been decimated. New "columns" and battle groups were quickly formed up and, along with some newly arrived reinforcements from England, were positioned within the defensive perimeter. Though the joke in Eighth Army was that a battle group or column was a brigade group that had been overrun by tanks, these improvised columns made up of the scattered remains of the brigade groups were motivated in battle.[119] One such column, Robcol, was made up of the remnants of the 11th Field Artillery Regiment as well as some British and Indian units from the 10th Indian Division and even some Gurkhas who had been with 18th Indian Division the day before. Demonstrating that small-unit cohesion was not always necessary for strong will, they fought firmly and played an important role in stopping the German advance on July 2.[120]

Newly arrived units showed their lack of training but also a good deal of morale and initiative.[121] The 23rd Armored Brigade had just arrived in the desert and had not received any training in desert warfare. That lack of training showed when they engaged the enemy head-on and lost a great number of their tanks.[122] Yet their continued willingness to engage the enemy in spite of the lack of confidence in leadership throughout the army indicates high morale and discipline. Additionally, those units that had been fighting for months continued to hold out against superior German equipment, even at high cost. The 4th Armored Brigade called their fight on July 2nd the "hardest day's fighting since the campaign began," but their continued fight in the face of heavy losses (17 Grant tanks, 19 Stuart tanks, and 3 Valentine tanks) helped to stop the last of Rommel's offensive actions in that sector.[123]

British forces were disorganized, exhausted, and underequipped by the time they fought the First Battle of El Alamein from June 30 to July 10, 1942. However, they surprised even their own commanders in their ability to stand against their opponent and hold Rommel's offensive long enough for his supplies to dwindle. Though there were severe problems with the armor and trust in high command was at an all-time low, British troops maintained their discipline and demonstrated morale in their overall commitment to winning the war, despite the conditions.

British forces paid attention to the politics of the war and identified its goals with the myths and symbols of their national identity. This led to high levels of will to fight against the Germans and Italians in North Africa. British soldiers demonstrated morale, discipline, and initiative even when their units were combined or put under new leadership. While the threat to the homeland was concerning, British soldiers saw the goals of the war as more extensive than preventing an invasion of Britain: they fought for freedom, democracy, and British honor as well. British perceptions of the threat posed by Germany was deeply informed by the myths and symbols of their national identity.

INDIA

Perceptions of the War Goals

In India, there were three approaches to explaining the war against Germany, none of which were widely accepted among the public or the military. The British government portrayed the fight as one to defend democracy, protect small nations, and champion the oppressed.[124] They suggested that anyone (particularly Congress) who opposed the war efforts was in fact aiding the fascists.[125] British propaganda attempted to paint Congress and other anti-war nationalists as appeasers and pro-fascist.[126] Beginning in October 1940, the British Indian government arrested hundreds of Congress party members who spoke against the war.[127]

At the same time, British officials found it difficult to explain the war aims to India. Viceroy Linlithgow argued that beyond resisting aggression, no statement of war aims was possible.[128] When nationalists of all stripes asked him to speak to Britain's intentions regarding India's status by the end of the war, he would only offer that "consultation" with all groups would take place.[129] When recruiting volunteers for the army, authorities specifically avoided making reference to patriotism, instead focusing on the practical material advantages available to soldiers—training, pay, and prestige.[130] In the summer of 1941 Churchill and Roosevelt issued the Atlantic Charter, and Indians wondered how a war fought for self-determination of all peoples was compatible with a war to preserve Britain's eastern empire.[131] When Indian leaders demanded clarification, Churchill claimed that the Atlantic Charter only applied to the countries under Nazi occupation and not to India or other British colonies. Even Indian leaders loyal to the British were outraged.[132]

The majority of nationalist groups—secular, Hindu, and Muslim—argued that the war was simply a war between competing empires. So long as India remained subject, Britain itself was no better than the fascist states.[133] They pointed to Britain's actions after World War I as proof that it was not truly interested in self-determination or democracy. As one editorial in the *National Front* noted, "Twenty-five years back India supported Britain's righteous war against Germany. Gandhiji and other leaders fell into the trap and rendered every help to the Government. India had to pay heavily for that error. The crawling orders of General Dyer [responsible for the Amritsar massacre], the Jallianwala Bagh massacres, and the Rowlatt Acts were the fruits reaped by India for the unforgettable error of 1914."[134] Congress, with its emphasis on socialism as an element of Indian identity, also pointed out that this was a war between two imperial-capitalist states, and they should be left to weaken capitalism between them.[135] Citing past unfulfilled promises of political power and an unwillingness on the part of the British to discuss India's postwar status, Congress urged Indians not to join the army or actively support the war effort. Though thousands of Indians still joined the army, many others participated in protest actions and statements. The British eventually made some 20,000 arrests.[136]

Muslim nationalist groups were also critical of the Muslim League for their cooperation with the British. They pointed out not only Britain's failure to keep its promises of democracy and self-rule but also its failure to uphold its promises to Muslims regarding the caliphate and the Ottoman Empire after World War I.[137] Each of these nationalist groups pointed to Britain's failure to live up to its own standards in regard to India as a reason to abstain from supporting the war effort. There was no reason of loyalty or pragmatism for aiding their colonizers. Most rejected supporting Britain's enemy but did not believe that they owed Britain their support either.[138]

There was a fringe of Indian nationalists that went so far as to actively support Britain's opponents. Subhas Chandra Bose lost his battle with Gandhi for power in Congress and fled to Germany, making an abortive attempt to build an Indian Army within the German Army to fight the British. When these efforts failed he went to Japan, where he had more success with Indian POWs in the Pacific (see chapter 4). In addition to attempted recruiting among Indian POWs, Bose began a series of radio broadcasts into India in February 1942.[139] Such active support of Britain's enemies was rare but did occur.

A minority of political leaders rejected Congress' skepticism and encouraged active participation in the war effort. They argued that Britain was fighting on the side of democracy, even if its own practices were not perfect, and supporting it in an hour of need might encourage Britain to increase India's own independence. An editorial in *The Leader*, written by one such nationalist, pointed to the dangers of abstention: "We may be told that if there is danger to any country, it is to England. But suppose England is defeated, is it then really believed that the socialists, or, for the matter of that, the Congress will be able to establish an independent government in India, or that Germany or Russia or Japan will let them do so."[140] While Congress tended to minimize any potential threat to India from Germany, some leaders argued that the threat of exchanging one unpleasant master for one much more tyrannical was real. Others, such as the Muslim League, saw Congress's abstention from government during the war as an opportunity to amass power and influence the British by providing active support for the war effort.[141] These pro-war politicians also pointed to the hypocrisy of Congress and leftist nationalists who claimed the war was simply one between two capitalist countries, all the while supporting the Soviet Union, which was allied with Germany until 1941.[142]

The Indian public closely followed the war and the politics surrounding it. The literate read detailed coverage of the war in English and vernacular newspapers, newspapers that were also read aloud to those who could not read. Additionally, radio broadcasts were played over loudspeakers from shops and at parks. Soldiers also sent news home in their letters, which were shared with the whole community.[143] Raj officials attempted to control the flow of information, but those efforts had limited effect.

Recruitment into the army was steady, with the army growing from 200,000 in 1939 to over 2 million by the end of the war.[144] British efforts to get Indians to help pay for the war through donations only backfired, leading to economic hardship and protest movements.[145] Yet victories in North Africa were celebrated with holidays and parades, while defeats spurred fear and even risked runs on banks.[146] The Indian public followed the war, allowed their sons to volunteer to fight it, but also protested it and refused cooperation.

Most of the Indian forces that initially fought in North Africa had been part of the military when the war began. These soldiers generally conformed to the "martial races" approach to military organization that the British had adopted long before, and the British made every effort to keep politics out of the Indian Army.[147] However, arguments regarding nationalism, the empire, and the goals of the war necessarily made their way into all aspects of life once the war began, even the Indian Army. Recruits were also part of the public discussions of the war goals, India's status, and support for or protest against war efforts. According to Khan, even the British officers in the Indian Army understood and sometimes shared their Indian colleagues' nationalist goals.[148]

As the army continued to expand, the British had to relax caste, ethnic, and educational eligibility in order to draw enough new soldiers.[149] The nationalists were more successful in deterring potential officers from enlisting, and those that did join as officers were more politically engaged than enlisted men, who tended to have less formal education.[150] Additionally, while the British expanded recruitment beyond the traditional martial races, those from "non-martial classes" were directed into noncombat roles. According to Raghavan, only 30,000 recruits from non-martial classes were employed in infantry battalions.[151] Officials also found it easier to recruit from regions that traditionally filled army quotas, where families had traditions of service in the army and knew the material and social benefits of that service.[152]

One group particularly noted in censorship reports as raising morale concerns throughout this period were the Anglo-Indian soldiers—men of both European and Indian descent. Censors note, "Something seems to be wrong with the Anglo-Indian element in the Force—as earlier summaries show a continual moan is going up from them in regard to unfair treatment. If promotion does not come quickly enough it is ascribed to as a difference in skins. If one gets in trouble, it is 'color' again."[153] Discriminated against for their race but without fitting into the British-ascribed categories for "Indian," Anglo-Indians faced significant barriers to integration into the army.

There were conflicts between men and officers of different classes as well as conflicts over the politics of the war.[154] Sikh units expressed concern about their homes, particularly in 1942 as talk of partition grew.[155] They were concerned about Punjab being included in Muslim Pakistan and the risk they believed that posed to their communities.[156] Sikhs in particular were a group that concerned the British, as they tended to be more politically active than non-Sikhs but had been a steady source of volunteers since 1857.[157] The Indian soldiers who went to North Africa to fight the Germans were thus heterogeneous in their own identities and in their perceptions of the goals of the war.

At the beginning of WWII, when the first Indian units were being sent to North Africa, India was beset by divisions—class and religious divisions over the identity and makeup of the nation and political divisions over the relationship between Indian nationalism and the war against Germany. There was not one single narrative that defined Indian identity, and neither was there a clear perception of whether the goals of the war against Germany—defeating fascism and totalitarianism but also defending British imperialism—were compatible with any Indian identity.

The Campaign

This section argues that Indian forces generally demonstrated high levels of will in North Africa, though there were a few exceptions indicating that such motivation was contingent on leadership and ethno-religious accommodations. I conclude that secondary cohesion supported Indian will to fight but that such cohesion could backfire if not situated in a larger commitment to the nation or state, making it a risky strategy to depend on for a mass army. This is borne out by the findings in chapters 4 and 5.

The Indian units that fought in East and North Africa were the prime units of the Indian Army. These units had been formed before the war began, benefiting from long training together, experienced officers (especially in managing issues of religion and ethnicity), and a sense of esprit de corps. As noted in chapter 2 when the Indian Army expanded to meet the demands of a global war, its British leaders were forced to expand their pool of potential recruits, thus lowering the educational requirements for officers and NCOs and introducing more ethnic and religious diversity into the ranks, drawing on groups without community ties to the army. But the units in North Africa were formed pre-war and thus had greater cohesion than their more recently enlisted comrades elsewhere. Mitigating against their success was the fact that none of these units were properly equipped in India and were slow to be outfitted (as were all units) once they arrived in North Africa.[158]

When Rommel began his first offensive in North Africa on April 3, 1941, the only Indian soldiers on the imperial lines were those in the 3rd Indian Motor Brigade, which was the first unit to take up a defensive position in the old Italian fortress at Mechili.[159] The 3rd Motor Brigade began the battle with a strong showing of discipline. According to one of the battery commanders with the unit, "Down through the years before the war whilst training in the Militia I had worked out exercises and maneuvers on sand tables and blackboards, but never did I imagine that such a huge force could be controlled as perfectly as was the 3rd Indian Motor Brigade and its attached troops on that morning."[160] The Indians had also acquired several small formations of British and Australian soldiers that they integrated into their defensive scheme.

Once ensconced in Mechili, the 3rd Indian Motor Brigade successfully repulsed an attack on April 6 by a small party of Germans sent out to mine the roads to the east.[161] They then turned down a request by the Germans to surrender, despite being surrounded by the advanced elements of Afrika Corps tank units and being outnumbered.[162] It was this engagement by the Indians that convinced the commanding generals that the Germans were in fact conducting a full-scale attack and that the Desert Army must withdraw to the east.[163] At this point the Indians were reinforced at Mechili by what was left of General Gambier-Perry's 2nd Armored Division headquarters.[164] The 3rd Indian Motor Brigade participated in several more actions defending Mechili as well as an attempted breakout, but soon the fortress was surrounded by German forces and General Gambier-Perry surrendered the garrison.

As described above, one squadron of the 3rd Indian Motor Brigade was left behind with an Australian and a British squadron as a rear guard. Demonstrating initiative, these three units decided to attempt one last breakout rather than surrender to the Germans. A number of them eventually made their way to Tobruk, where they joined that garrison for the remainder of the spring and summer.¹⁶⁵ Major Rajendrasinhji's squadron, having trained and fought together for some time and led by an experienced officer, exercised the discipline and initiative to break out through German lines rather than surrender with the rest of the garrison. The major received the Distinguished Service Order, the first ever awarded to an Indian officer.¹⁶⁶ Small-unit cohesion, in particular strong leadership, led to high will to fight in this instance. It should be noted, though, that three units of different nationalities who had been together for only a week were able to coordinate this action together, suggesting that time together is not a necessary element of cohesion.

As the rest of the Desert Army withdrew back to the Egyptian frontier, the escaped members of 3rd Indian Motor Brigade as well as the 18th Indian Cavalry Regiment, who had arrived in North Africa just in time to join the retreat, took up residence in Tobruk. When the siege of Tobruk began, there were 15,000 Australians, 7,500 British, and some 500 Indian soldiers in the fortress.¹⁶⁷ The Indian soldiers had rifles, Bren guns, and anti-tank rifles but no armor.¹⁶⁸ Though there were very few of them, the Indian soldiers actively participated in the defense of Tobruk, including the night raids for which the garrison became famous.¹⁶⁹ The other soldiers in Tobruk were impressed enough by the Indian troops' performance to write home about it:

> The Indians are good fighters. When they go on patrol they go off for a few days into enemy lines and come back with a bag full of ears and plenty of information. Apparently they use a knife, and every jerry they bump off they swipe his ears as proof. Often they pierce right back to the enemy HQ and come back with badges of the big noses got off the tunics they wore at the moment of their finale.¹⁷⁰

I have found no evidence that Indian soldiers mutilated their opponents in this way, but the letter suggests both the racism that colored British views of Indian soldiers and the Indians' reputation for initiative in Tobruk.

General Morshead, as he prepared to leave Tobruk with the rest of the 9th Australian Division in October, wrote,

> It is appropriate to mention here the ascendency which the 18th Cavalry Regiment had always maintained over the enemy. This unit, untrained for such duties, took up its allotted position in the front line as infantry practically from the time the defenses were first occupied and remained there continuously until the time of their embarkation at the end of August. By *their fighting spirit*, venturesomeness, and constant alertness, these stalwart Indians succeeded in defending a

very wide front for a long period, throughout which the enemy was made to feel and fear their presence.[171]

Both commanders and soldiers in Tobruk recognized the will to fight of the Indian soldiers with whom they fought.

The soldiers of 4th and 11th Indian Infantry Divisions were an active part of the summer campaigns, Brevity and Battleaxe, though the campaigns themselves were disappointing.[172] War diaries note that morale was high, despite difficult fighting.[173]

At least one Indian unit did, however, experience conflict over religious issues while on its way to North Africa. Shipboard, the 11th Sikh Regiment was given the order that each soldier must carry his steel helmet with him at all times, though they were not ordered to wear them. The helmets could not be worn with the Sikh turban, and the soldiers were concerned that should they be ordered to wear them they would have to take off their turbans, a violation of their religious practices. Indeed, "Some of them professed religious scruples which prevented them even picking them [the helmets] up."[174] There was much dissension over the issue, with some soldiers saying they would refuse to carry the helmet even with a direct order. Though the officers suspected political instigation by men who had been transferred in from another unit, they eventually decided not to risk soldiers defying orders and collected the steel helmets, saying they would not enforce the order to carry them.[175] Elsewhere in the empire, the order to carry steel helmets led to outright mutiny among Sikh units—such an extreme was avoided in this case by officers who decided to ignore high command.[176]

The incident demonstrates the importance of officers who know and respect their men, as well as the potential for conflict over ethno-religious differences in the Indian Army. For these soldiers, religious identity came before their loyalty to the Indian Army. This breakdown in discipline came shortly after the end of the civil disobedience campaign in India, where over 7,000 had been convicted of violating wartime laws.[177] Additionally, relations between Sikhs and the British Raj were deteriorating as the British seemed to be open to the idea of partition, which Sikhs generally opposed.[178] Sikh units were sensitive to any signs of disrespect to their identity from the British. This was particularly concerning for the British, as Sikhs made up 17.51 percent of the Indian Army in 1939 despite accounting for just 1.4 percent of the population.[179]

After a period of quiet from July to November, during which General Auchinleck took command and the Desert Army became Eighth Army, Operation Crusader got under way in an effort to push the Germans back into Libya and relieve Tobruk. Units from all three nationalities were also resupplied over this period.[180] Both the 11th and 5th Indian Infantry Brigades and the 7th Brigade Group, of the 4th Indian Infantry Division, were once again in battle and exhibited discipline and good morale.[181] On November 21, 4/16 Punjab Regiment (of the 7th Brigade Group) participated in an offensive against a German-held position, sustaining many casualties but managing to capture and hold a few positions, taking 1,500 prisoners as well as guns and equipment.[182] After several days' delay, they captured the entire position (after the German troops had withdrawn, leaving the final defense to the Italians). While 4/16 Punjab did not capture their position as quickly as the Royal Sussex, a British unit, had captured a similar nearby position, 4/16 Punjab did not have a full complement of tanks and initially came up against German troops, as opposed to the Italian troops facing the British unit.[183] Despite these difficulties, the 4/16 Punjab demonstrated discipline in overcoming them.

The 4th Division saw very difficult fighting at the end of the battle; on December 15 the German 8th Panzer Regiment overran two of its battalions, killing or capturing over 1,000 men.[184] But despite the Indian troops' losses, they fought with discipline. They were able to destroy half of the tanks the 8th Panzer Regiment sent against them and delay the Germans long enough for 4th Armored Brigade (a British unit) to move into the Panzer Regiment's rear.[185] The Eighth Army succeeded in pushing the Germans to the east of Tobruk and relieving that fortress.

On the same day that the Germans pulled back from Tobruk, the Japanese attacked Pearl Harbor and Malaya, marking the beginning of a direct military threat to India.[186] Interestingly, it does not appear that Indian soldiers reacted strongly to the threat Japan posed to their homeland. After December 7, Australian soldiers were very concerned for their homes, and many wanted to return home to join the fight against Japan (see below). I found no evidence that Indian soldiers saw the Japanese as a threat to their homeland that they wished to fight off themselves. While British officials discussed the effect that the war with Japan had on Australian morale in North Africa, I found no mention of a similar concern over the morale of Indian forces.

When in January 1942 the Germans attacked, they succeeded in pushing imperial forces back to the Gazala line at nearly the same time that the imperial forces in Singapore surrendered to the Japanese.[187] Another long pause in major activities occurred from February to May 1942, during which both sides re-equipped and absorbed reinforcements. Expansion of the Indian Army at this point was well under way, but most of the new units were staying in India to protect against potential Japanese attack and prepare for eventual counterattack against Japanese forces across the border in Burma. There were no major reinforcements of Indian units during this period.

When Rommel attacked the Gazala line in June 1942, the Eighth Army did not at first realize that it was a full-scale offensive. As a result, the 3rd Indian Motor Brigade, which had begun its fight over a year earlier in Mechili, was surprised by a German Panzer regiment while at their breakfast.[188] Despite being attacked while making their morning tea, the brigade still engaged the attacking Germans in a "fierce fight."[189] It was only once this brigade was overrun and lost over 200 soldiers and a number of officers that Auchinleck and his commanders grasped the gravity of the situation. Yet even when they were overrun, there was no surrender. The unit demonstrated discipline by retreating as ordered and rallied at predetermined points in order to attempt to rejoin the rest of the Eighth Army. They finally succeeded in doing so that afternoon, joining the 29th Indian Infantry Brigade, which had come out to meet them, and drove off the pursuing tanks.[190]

Indian units did demonstrate some indiscipline as the Germans broke through the Gazala line. According to the official history, on June 6 the 21st Panzer Division overran a defensive position in the area known as "the Cauldron," and the forces there scattered. Unlike prior retreats where Indian troops maintained their order and even demonstrated initiative in breaking out of encirclements, this time chaos reigned. As the headquarters for the 9th and 10th Indian Brigades were attacked, "communication and control broke down completely."[191] The 9th and 10th Indian Infantry Brigades dispersed in confusion and were unable to regroup and rejoin the battle. All this while the British 22nd Armoured Brigade failed to intervene; they had been told they were not responsible for the infantry.[192]

When the imperial forces were able to slow their retreat and regroup at the Alamein line, a number of new units or columns were formed out of the remnants of units that had broken in the retreat.[193] One particular unit was made up of members of the 11th Field Artillery Regiment, members of the 10th Indian Division, and some Gurkhas of the 18th Indian Brigade. This unit fought firmly and played a crucial role in halting Rommel's advance on July 2.[194]

The Gurkhas from the 18th Indian Brigade were some of the few to have escaped from the fighting the day before. The 18th Indian Brigade was defending a position near Deir el Shein without aid of armor.[195] This unit bore the brunt of the renewed German thrust when on July 1 the veteran 15th and 21st Panzer Divisions attacked. The 18th fought hard and held the Germans for a full day longer than Rommel had anticipated. Despite the lack of armor, and even after they were told that the 1st Armored Division would not be able to reinforce them, the 18th fought on.[196] The Sikh unit within the 18th even managed to break out and rejoin the brigade further east after being surrounded by German tanks.[197] The 18th Indian Brigade displayed morale, discipline, and initiative in a fierce battle against a better-armed opponent. The delay imposed on the German forces here allowed the remaining imperial forces to redeploy and resupply and was key in blunting what Rommel had intended to be his final thrust into Egypt.[198]

Toward the end of the battle at Alamein, Indian forces participated in counterattacks on German positions with mixed success. The 4th Rajputana Rifles broke up under heavy fire attempting to take a German position. They were reassembled and sent to take an Italian post instead, where they had more success.[199] It was on July 14 that the armor and infantry units finally began to work together in a real combined effort. The 2nd Armoured Brigade and the 5th Indian Brigade integrated tanks and infantry, resulting in one of the most significant victories of the fighting at Alamein.[200]

By the summer of 1942, the political situation in India was deteriorating. Congress had rejected a proposal aimed at settling India's constitutional status (described in more detail in chapter 5) and was debating instituting Quit India demonstrations throughout the summer. Gandhi launched the demonstrations on August 8, 1942, and though he and the other Congress leaders were arrested the next day, rioting and political activism continued for weeks afterward. One Indian civilian wrote to an Indian officer in North Africa, "Everyone is waiting for the Congress to start their Satyagraha. I feel the Congress people will stand up to it, but what use is it unless they can get the peasants and workers to participate. Anyways, I hope they succeed."[201] Censorship reports indicate that news of the debate reached Indian soldiers through mail from home. Despite risk of punishment, some Indian soldiers in August indicated support for the movement. According to the report, "A Havildar of 2 B.S.D. on whom retribution will shortly fall, in a letter for Foresepur says: 'Please inform Congress to persuade people not to join the Government Service, and if they do, then they will all be victims of this war.'"[202] Another letter, presumably censored extensively before being passed on, described the writer's attendance at the All-India Congress Committee meeting in Bombay, hearing speeches from Gandhi, Nehru, and others. He wrote, "It was an inspiring, reverberating appeal and calling to India. Nay to the world and particularly to the United Nations of America, Britain, China and Russia."[203] Anti-war nationalist politics was beginning to trickle into Indian Army units, even at the risk of imperial "retribution."

There were, however, other letters encouraging family to join the army and expressing hopes of being able to remain in the service after the war was over. The army offered financial and career opportunities that appealed even in a time of war. One Indian soldier wrote "those who are abroad are the true warriors.... So long as there is life in our bodies, we will fight and hit the enemy vigorously."[204] There was significant variation in the way that Indian soldiers responded to nationalist mobilization in India during this period, especially among the long-serving soldiers stationed in North Africa.

Indian forces in North Africa demonstrated generally high levels of will to fight throughout the campaign. These units had been together for some time, had received extensive training—though not necessarily for desert warfare—and had officers experienced with the Indian Army in general and their units in particular. However, there were occasional conflicts over ethno-religious issues such as the order to carry steel helmets. While such conflicts resulted in outright mutiny in the Pacific, experienced officers in North Africa were able to avert escalation and accommodate their soldiers' needs. Unit cohesion without some sense of commonality with leadership and cause could weaken or even work against combat motivation under certain circumstances.

AUSTRALIA

Perceptions of the War Goals

Australia joined WWII within one hour of the United Kingdom's declaration of war. While prime ministers in Canada and South Africa put the question of joining Great Britain's war to their parliaments for a vote, Australia considered the decision to be automatic. There was no political contention about joining Britain's faraway war against Nazi Germany.[205]

Even the growing evidence that Great Britain did not have the military capability to ensure Australia's security should war erupt in the Pacific did not cause the Australians to take their foreign and military policy into their own hands.[206] In 1937 the Labor government suggested shifting from building ships for Britain's Singapore strategy plan toward building an air force to defend the Australian homeland. Australians viewed this as a sign of Labor's desire to "go it alone" and duly punished the party in the elections that year.[207]

In 1939 the war that lay before the empire was still a European affair. Germany was expanding across Europe, and Great Britain had drawn a line in the sand at Poland. To that point Britain itself faced no direct threat, and Australia was thousands of miles from the nearest German soldier. Yet the public accepted Prime Minister Menzies's statement at face value: "Britain has declared war . . . and as a result, Australia is also at war."[208] This is not to say that there was not political argument surrounding the war in Australia. Though declaring war and joining Great Britain were taken as a given, the actions and policies required for this were debated for the rest of the war.

Some political leaders expressed concern for Australia's security should they send large numbers of men abroad to aid Great Britain.[209] Japan posed a growing threat within striking distance of the Australian homeland, and the British were clear that their first priority lay in defeating Germany and defending Britain itself. These political leaders—primarily made up of opposition parliamentarians—were concerned that Japan would see a weakened Australia and take the opportunity to infringe on Australian interests or even Australian territory. At the end of 1941 this was indeed what happened. However, commitment to the empire won out, and thousands of Australian soldiers would fight the first years of the war far from home.

Australians viewed the war as one to defend a democratic way of life. When Menzies sought extensive executive powers in order to put the country on a war footing, there was concern throughout the country. "For some, there was a dangerous contradiction in the restrictions necessary for effective prosecution of the war and the prime objective for which the war was being fought, namely that of preserving a way of life in which the rights of individuals might still count."[210] Changes to the law and the economy were slow, and aimed to preserve the democratic and individual rights that Australians believed they were fighting to defend.[211] In fact, in October 1941, 43 percent of Australians surveyed by Australian Public Opinion Polls stated that they were dissatisfied with the way the current government was conducting the war (they were not offered an opportunity to explain why).[212] Forty-two percent said that Australia was doing enough toward winning the war, while another 42 percent stated that it was not.[213] The lack of full mobilization at home was a bone of contention with the Australian troops fighting abroad; they felt that those at home were less than committed to their success.[214]

Australians initially engaged Italian forces in Libya, facing German forces when Rommel entered the fray in March 1941. The imperial forces viewed their Italian opponents as less than worthy. Even before they actually engaged the Italian forces in combat, the Australians especially viewed them as having "no 'heart,'" as weak and in a hurry to surrender.[215] Throughout the remainder of their time in North Africa, Australian soldiers dismissed the fighting abilities of Italian soldiers and viewed them as a sideshow to the greater enemy—the Germans.[216] Some Australian soldiers even noted that the Italians were not committed to Mussolini's cause and had no desire to fight at all.[217]

The Germans, on the other hand, were viewed with great respect but as the clear enemy. It is important to note, especially given Australian views of the Japanese later in the war, that the Australians rarely dehumanized their German opponents. The German was the enemy and he fought for a terrible cause, but he was also a good soldier and an honorable opponent. For example, one soldier writing from Tobruk noted:

> The German is a worthy opponent and in this campaign at least he is a clean and fair fighter ... I have yet to see a German who resorts to low and mean subterfuge. What a pity they are so blindly fanatical in their political creed—what a pity they shun God and the Son of God—what a pity their courage, their talents, their thoroughness and ingenuity merely go towards building up a clay god of War.[218]

German behavior was interpreted through Australian myths of whiteness and democratic norms. German actions in war were honorable, but their cause was shameful. Australian soldiers clearly believed that the enemy's motivation was directly related to their political beliefs. Additionally, it was not dehumanization of the enemy that allowed the Australians to justify killing them. Rather, the goals for which the enemy fought were at odds with Australian identity and thus justified the violence.

Australians viewed the war against Germany and Italy as one to defend the empire and democracy against an aggressive, totalitarian foe. The clarity of the threat to the empire was such that Australians did not even require a political debate over joining the war. There was little concern for the direct safety of Australia; in fact, there was concern that actively joining the fight in North Africa might actually weaken Australia in the face of the real threat—Japan.[219] In the end, the risk of weakness relative to Japan was considered acceptable in order to live up to Australia's commitment to the empire and its principles.

Australian attitudes toward the war with Germany shifted with the Japanese attacks on Pearl Harbor and Singapore. There was a strong concern for the safety of the homeland and a desire to be home to help defend it.[220] Certainly the Japanese threat to their families at home led many Australian soldiers to view the war in North Africa as one of secondary importance. However, as I argue below, Australian soldiers who did remain in North Africa continued to view themselves as participating in a worthwhile fight that needed to be won.²²¹ They continued to view their role in North Africa as important and worthy of their sacrifice, even as they worried about their homeland.

The Campaign

Australian forces began their combat activities in the Second World War in the North African desert. They displayed high levels of will to fight throughout their time in North Africa. Australian troops showed high levels of discipline, morale, and initiative in the numerous battles in which they took part.

The 9th Australian Infantry Division arrived at the front following Operation Compass, which had pushed the Italians out of the Egyptian frontier and to the far western edge of Libya. This was the 9th Division's first time on the front lines. They arrived at the end of March to take over the positions of units that had gone to Greece.²²² The Australians' first encounter with the Germans came on April 4 outside Benghazi where they had taken up defensive positions to prevent the German troops who had just retaken the city from proceeding on toward Egypt should they decide to continue eastward.²²³

The Australian soldiers of 2/13th Battalion, "D" and "A" Companies, who first engaged the Germans did so with discipline, morale, and even some examples of initiative. They were outgunned yet held off their attackers for some time before giving ground.²²⁴ When they did retreat, they did so as a unit and remained in the battle area waiting for reinforcements. They even replied with grenades when a German patrol called into the tank ditch in which they were hiding, offering to accept their surrender.²²⁵ This small group of Australians spent the next few weeks attempting to catch up to its retreating headquarters.

The next Australian engagement came for two units of Australian antitank guns under the command of a British artillery squadron. Having been caught away from 9th Division when the Germans attacked, the Australians simply joined up with a British unit and an Indian unit that had also come under British command.²²⁶ Their lack of training together or sense of common unit history do not seem to have decreased their will.

The Australians were among many trying to break out from Mechili (as described above), having been cut off by the German attack. On April 8 the breakout failed, and while some units surrendered along with their commander, General Gambier-Perry, the Australians and their British and Indian comrades were able to break out and make for their own lines.²²⁷ Unlike the Indian and British soldiers in this group, the Australians were not organized to develop small-unit cohesion and had little experience together. But that lack of small-unit cohesion did not impair will.

The Desert Army as a whole was making a fairly quick retreat in the face of the German attack, and the Australians' next engagement was a long one. On April 11, the last of the 9th Australian Division made its way into Tobruk fortress, where they remained until October. The roughly 30,000 troops there were primarily Australian, joined by an odd assortment of several thousand British and several hundred Indian troops described above.

Constant retreat combined with little to no engagement for most of the Desert Army meant that morale suffered. Tobruk's access to a port led some to believe that a second Dunkirk was in the making, a continuation of their retreat by sea. On April 11, the Australian General Morshead, fortress commander, made it clear that would not be the case.[228] Morshead's strong leadership, combined with his orders for aggressive patrolling of the perimeter convinced all the troops in Tobruk—Australian, British, and Indian—that here they would take a stand against the enemy and morale improved.[229]

It was at Tobruk that the Germans met with their first major defeat in the war. Rommel was surprised and angered by the failure of his blitzkrieg strategy to drive the Australians out of their defenses as it had the armies of Europe.[230] The frontier posts bore the brunt of the German offensive, facing panzer tanks, infantry, and dive-bombing. That combination had proven successful against other armies, but the Australians stood firm, exercising strong discipline in waiting out the attack in order to engage the infantry.[231] For most of them, these early battles were their first encounters with German tanks, yet the panzers did not incite the overwhelming fear that they had in Europe. Fighting often descended into combat with bayonets in the frontier posts as they changed hands several times.[232]

The injury or death of unit leaders did not prevent Australian units from carrying out their mission. One unit was sent to mop up a German tank battalion that had been broken up after having breached the perimeter. Their leader, Sergeant McElroy, was shot by a sniper as he led them toward a German-occupied building. This served only to anger the Australians, who proceeded to charge the building, hurling grenades. As the Germans fled the explosions, a number surrendered. Those who did not were bayonetted.[233] Though dedicated to their leaders, especially the NCOs, Australian soldiers were consistently able to carry out their mission even without their original leadership.

Australian soldiers wrote home playing down the poor conditions, telling family and friends to "swallow such reports with a 'pinch of salt.'" Moreover, censors noted, "Australians gladly suffer 'Hell' so that such tribulations may not visit 'Aussie.'"[234] Morale was good as AIF soldiers besieged by the Germans in Tobruk believed they were protecting their home on the other side of the world.

During this same period, the Germans gained an appreciation for the willingness of Australian soldiers to make risky counterattacks. One German soldier described the siege of Tobruk from his perspective outside the perimeter: "Suddenly a wireless message. Enemy infantry are on the attack. Are they crazy? No, it's actually true. Two Australian companies are leaving their lorries and forming skirmishing lines."[235] Axis forces were often surprised in their trenches by Australian patrols and small counteroffensives. These risky patrols fulfilled a standing order to conduct an "aggressive defense" but were planned and executed at the unit level. Soldiers could have gone out and avoided contact with the enemy, yet consistently throughout the siege, Australian soldiers sought out and destroyed enemy positions, captured prisoners for interrogation, and bayonetted those less cooperative.[236] Australians in Tobruk demonstrated strong discipline and initiative, despite difficult conditions.

Despite complaints about Australians receiving all the attention, British soldiers did respect their comrades' achievements—even in letters home. One soldier described the Australians thus: "Another Aussie incident happened last night. Jerry had brought in a new mortar and had shelled these Aussies. They said 'we'll have that tonight' and the following morning a new mortar and ammunition was in the Aussie pit and they didn't lose a man taking it."[237] Australian soldiers returned the compliment, writing home: "But for those Pommy Boys of the R.H.A. we would have been wiped out, and any Aussie who has been in Tobruk will take his hat off to those English Boys. You've got to live with them for a while to understand them and for them to understand me, but underneath we're all the same."[238] Frustration at the press coverage of their battles was a constant in letters home, both British and Australian, but it does not seem to have affected their ability to work together effectively and appreciate one another's military performance. Moreover, there was a sense of similarity and connectedness with the British that Australian soldiers noticed, despite the somewhat derogatory terminology.

Morale was still strong throughout this summer, despite general irritation with the conditions. According to the official censorship reports, Australian troops in particular mocked the German propaganda efforts, mailing the flyers that were dropped over Tobruk home as souvenirs.[239] The strongest complaints in the mail home regard the inconsistency of mail delivery and a failure to deliver "comfort funds" packages.[240] There are some signs of conflict between Australian and British troops, as the AIF was receiving much press over their exploits in Tobruk while the British soldiers believed their accomplishments were not being acknowledged.[241]

The Australians remained in Tobruk until their government insisted on their relief in September and October. When Australia sent the AIF abroad, it had done so on the condition that Australian forces would fight as a single corps. The Australians had permitted that corps to be split when Wavell sent the 6th Division to Greece while the 9th and 7th remained in North Africa. After the disastrous battles there and the retreat from Crete, the Australian government insisted that their forces be reunited, which meant relieving the 9th Division in Tobruk and replacing them with British troops.[242] Churchill, the British chiefs of staff, and General Auchinleck all objected on military grounds, but the Australian government was insistent. Most of the British and Indian troops would remain for several more months while the Australian soldiers were replaced with British and Polish forces.

Interestingly, the censorship reports indicate that politics was a topic of some discussion, despite some historians' claims to the contrary.[243] After the German invasion of the Soviet Union in June 1941, numerous letters home discussed the state of that front in the war against Hitler. The censorship report notes that, "the idea of 'breaking in somewhere' whilst Germany is preoccupied on the vast Eastern front is mooted in many letters. Hundreds of men out here think now is the time to attack."[244] This shows that the Australian troops paid attention to the politics of changing alliances as well as the military implications, and hoped for a chance to become more active on their own front.

Additionally, Australians were concerned with the political goals of the war. One soldier, having visited England before being shipped to North Africa, compared the British people's understanding of the war goals with his own people's and found the Australians' lacking. "The government in Australia seems to be doing a fairly good job but there is one thing lacking in their policy which is of vital importance and which is receiving a lot of attention in England. This is the statement of war aims and the announcement of a tangible outline of the new order that must follow this war if civilization is to make any progress."[245] This particular soldier's concern notwithstanding, the Australian military did attempt to educate its soldiers on the politics of the war through some of its military publications. Beginning in December 1941 the army distributed the troop journal, *SALT*. It published an article on the failure of British and French appeasement, suggesting that their desire to avoid war combined with their tendency to choose the path of least resistance led to inaction and failure to stop Hitler.[246] *SALT* also published a series of articles on what the Australian government hoped to achieve in a postwar world: "What, then, do we want our post-war world to be like? Here I can speak only for myself, but I think a great many people will agree with me. First, we want to get back to democracy in its truest sense. We want a society in which the people govern themselves through institutions in which they have faith and confidence, and these can only be based on knowledge and understanding." [247] Australian soldiers and their government thought that an understanding of the goals of the war was important for motivating the troops and the home front. They particularly emphasized the goal of protecting and reinforcing democracy—a goal related to Australian identity as an outpost of democracy in the Pacific.

In addition to the progress of the war, domestic politics also received much comment from Australian soldiers—often negative. One censorship report noted, "Many members of the AIF in the Middle East follow the course of domestic politics with keenness and often disappointment. They think that the political parties in Australia should do less wrangling and get on with the job of winning the war."[248] Australians who had stopped off in England before being sent to North Africa contrasted the war effort in that country with the war effort back home and found their own people's efforts to be lacking.[249] The Australian government never formed a national unity government as the British did, and the changes in government as well as civilian actions such as factory strikes caused frustration among the troops in North Africa, contradicting the expectations of democratic effectiveness theory.

Australian soldiers faced a unique situation beginning in December 1941, when the Japanese attacked Pearl Harbor and Singapore and began expanding toward Australia. By this time the majority of the Australians were in Palestine training or garrisoning Syria. In early December, as the Japanese were scoring successes throughout the Pacific, censors reported that in Australian letters home:

> In spite of the Japanese successes in the Far East and the homesickness referred to above, the morale of the troops continues to be excellent. . . . The close cooperation between the Democratic Powers is commented on most favorably by many and is regarded as a happy augury *not merely for victory but for the creation of a post-war world where democratic ideals will always remain supreme. That this is a just struggle against a diametrically opposed ideal and not just a fight into which anyone can join is more and more realized.* The Japanese successes in the Far East, though regarded as serious, are not considered to have any lasting importance. There is a deep conviction that they

are only temporary and that the vast forces which the USA and Britain will shortly be able to bring against them will prevail. There have been few signs of uneasiness about the safety of Australia and a few demands to be sent back there, but there is definitely a desire to "get at somebody, preferably the yellow bellies."[250]

This assessment of Australian letters home indicates that Australian soldiers had a clear grasp of the wider context of the war and were committed to the political goal of protecting democracy in line with their national identity. It also indicates that the soldiers, though concerned about their homeland and even expressive of racist attitudes about their new enemy, saw the war against Japan as part of the wider fight in which they too were participating in North Africa.

This did not stop many soldiers from wishing to return home to defend their families, especially as the Japanese seemed to be unstoppable in January and February.[251] Indeed, some soldiers perceived the war with Japan to be more important than their duties in North Africa—which at the time consisted of garrison duty in Syria. "There are thousands of anxious Aussies over here who want to be home doing their bit for what seems a more worthy cause. We would lick the hides of the Japs."[252] Much of this frustration was tied to the fact that the 9th Division was not actively participating in any fight at all in North Africa: "A number of personnel from the Australian Imperial Force in Palestine and Syria firmly believe that they are being wasted in these countries and should be in action in the Far East."[253]

The sense of waste is not surprising, given that Australian troops were not conducting combat operations in Palestine or Syria. In fact, frustration at their lack of action in Syria had been building even before Japan's entry into the war, as noted by the censors' reports.[254] Additionally, race played a role in Australians' assessments of their situation, with a number of letters using racial slurs for Arabs and describing the region as "rotten."[255] On the other hand, one soldier who was in action expressed satisfaction and a sense of patriotism in having a chance to fight "Jerry," despite some losses to his own unit. "Our boys did a marvellous job and it makes me proud to be an Australian."[256] Another emphasized that until imperial forces went back on the offensive, the empire's prestige and confidence would continue to fall—"attack we must soon, lest we lose faith in ourselves and our destiny."[257] Australians were concerned about their homes and frustrated with their lack of action in North Africa throughout February and March.

Yet even as Japan was defeating their countrymen in Malaya (see chapter 4), Australian troops still viewed themselves as part of a wider imperial war effort. The censorship report for February 11–17, 1942, quotes one soldier who criticized the Australian prime minister: "Most of the lads are pretty sore with Mr. Curtin, and the way he said that England has almost left Australia to its doom. I can understand him taking precautions, but he doesn't need to panic. We think the English people have done a wonderful job."[258] Soldiers of the 9th Division expressed understanding that they were still needed in North Africa.[259] A March 1942 report notes, "The general feeling of the troops is one of disappointment at being still in the Middle East while events of such importance are taking place around Australia. They find consolation, however, in the hope that they are to have a job of almost equal importance."[260] Australians would have liked to be in the fight against Japan but recognized the fight against Germany as an important part of the overall war effort.

By March 1942 the 6th and 7th Divisions had arrived back in Australia at the request of the Australian government, but because of shipping shortages the 9th Division remained in North Africa. Real problems with morale and attitude among the 9th Division did arise in April when a newspaper in Australia published an article that claimed that those Australians still in North Africa had "volunteered" to remain, which was not true.[261] Given that this report came at the same time that American troops were streaming into Australia, reactions were unpleasant: "We are fairly quiet here and more or less resigned, though the report that we volunteered to stop here is a lot of tripe, our sympathies at home. Several of the lads have had letters from wives and girlfriends telling of the lovely American lads they have met. Their remarks were distinctly Aussie and very obscene."[262] Soldiers did not want their family to think that they preferred to stay on guard duty in seemingly quiet North Africa while the Americans defended their homes.

In this same period, an Australian officer wrote home:

> Let us be dinkum Aussies enough to look at the facts fairly and realise that it's England that stands at the head of our column: it's England who first stood up and said to the Hun stop, and when the Hun didn't stop, took the huge risk of declaring war because it was the way of the brave. Now when the British Empire is up against it, this is the time for every Britishor throughout the world to take his courage in his hand and say, "this is today's link with the past, with deeds of Drake and Nelson."[263]

Australian soldiers worried about home, were concerned about Japanese victories in the Pacific, and wondered if they were making a difference. But they also saw themselves as part of the British Empire and connected with its history of defending the world against aggressors, invoking British and Australian imperial myths to make the case.

When Rommel again took the offensive in North Africa in late May 1942, driving the British out of the western desert and back into the Egyptian frontier, the 9th Australian Infantry Division returned to battle. It went first to Alexandria and moved up to the Alamein line to join the First Battle of Alamein on July 3, 1942.[264]

Having received reinforcements from Australia in November and December 1941 while in Syria, the Australian units were among the few who were up to full strength when they went into battle.[265] However, they lacked any transportation or mechanized support, so they had to rely on British motorized units, tanks, and artillery when they went into battle. They also lacked training for mobile action; most recently they had been on garrison duty in Syria and before that had conducted a fixed defense at Tobruk.[266] Nevertheless, Australian soldiers were eager to return to battle. According to censorship reports from June 1942, "The possibility of a 'move to the western desert' has improved morale. Those who have already seen service are not particularly desirous of returning but state they would go willingly if ordered to do so. Confidence in our final victory is never doubted, and the troops' paramount desire is to get into some action and help to bring the war to an end."[267] Those who had participated in the defense of Tobruk were also angered by its loss (under South African garrison on June 21, 1942) and interested in reclaiming it for the empire.[268]

The Australian forces exhibited high levels of will in the battle at Alamein. Australians participated in counterattacks on German defensive positions beginning on July 10 until the battle came to a close at the end of the month, demonstrating discipline, initiative, and high morale. Initial counterattacks by Australian units resulted in an important salient into the enemy line and a large number of mostly Italian prisoners.[269] Some scholars have pointed to these battles of July 10 as the turning point of the campaign, when the imperial forces shifted from defense to offense.[270] After the Australians' first successes, Rommel shifted his German troops to the north to meet them, and the Australians faced a tougher opponent.

The salient came under attack on July 15, and the Australian infantry held off German infantry and tanks.[271] The next day they sought to expand the salient further but quickly encountered artillery fire and much more resistance from the German troops than the Italians had put up. Here the Australian soldiers demonstrated initiative, charging machine-gun posts and even engaging in bayonet charges reminiscent of actions at Tobruk.[272]

The position was captured but without artillery and tank support could not be held, and the Australians withdrew, making their way back to rear lines.

After July 16 the Germans retook the offensive, and the Eighth Army once again found itself on the defensive. Refusing to abandon their position in the face of superior equipment, Australian infantry actively defended their posts against tanks. The Australians lay down in their slit trenches and allowed the tanks to drive over them, and then engaged the infantry that followed, leaving the artillery in the rear to take care of the tanks. In some cases, if the German infantry did not follow quickly enough, the Australians engaged the tanks themselves, using grenades.[273]

Finally, Australian units were able to maintain discipline and continue with their mission even if they lost leaders. As at Tobruk, new leaders were able to exercise command and lead their troops in battle should the necessity arise. During one assault on German defenses, the 2/24th Battalion lost all but one of its officers, and he had been away for some time. "Lt. Austin who, having rejoined the battalion only the day before, found himself in charge, scarcely knowing the men."[274] Nevertheless, he was able to exercise command over a number of platoons and coordinate their continued attack. Leaders with little leadership training or who were unfamiliar with the unit were not an impediment in the Australian Army.

Australian soldiers were motivated to fight by their commitment to defending Australia, which they understood as being an integral part of the British Empire and thus at risk because Britain was at risk. They also considered themselves to be fighting for democracy, even as they criticized their own democratically elected government on occasion. The myths and symbols of their common national identity motivated them to fight against an enemy far from their own homeland.

CONCLUSION

All three nationalities fighting the Germans in North Africa had high will to fight. British, Indian, and Australian soldiers demonstrated high morale, good discipline, and initiative in battle. National identity theory explains British and Australian motivation, as well as their sense of fighting to defend democracy, their way of life, and their homelands. Small-unit cohesion also helped motivate the British and especially the Indian soldiers in North Africa, though reliance on that cohesion did pose some potential dangers. When Indian units' identity clashed with that of their British leaders, small-unit cohesion enabled the unit to stand against their leadership. So when the steel helmet order was issued, the entirety of the 11th Sikh Regiment could oppose the order. In this case the officers understood their men's objections and chose to ignore the order themselves rather than risk mutiny, but mutiny did occur over that same order in Sikh units in the Pacific. Small-unit cohesion can work against the larger military organization in certain circumstances. Having a common set of myths and symbols through which to filter and explain policies and war goals helps to avoid such clashes.

NOTES

1. Hastings, *Inferno*, 105.
2. Jackson, *The Battle for North Africa*, 28
3. Jackson, *The Battle for North Africa*, 34; Pitt, *The Crucible of War*, 32.
4. Jackson, *The Battle for North Africa*, 94.
5. For more on this topic see Jackson, *The Battle for North Africa*, 160; Barr, *Pendulum of War*, 49; Carver, *Dilemmas of the Desert War*.
6. Bharucha, *The North African Campaign*, 132.
7. Barr, 157.
8. Weight, *Patriots*, 48.
9. Churchill, *His Complete Speeches*, 6380.
10. Churchill, *His Complete Speeches*, 6476.
11. Weight, 40 and 65.
12. Greenwood, *Why We Fight*, 28.
13. Greenwood, 122.
14. Gallup, *Gallup International Public Opinion Polls*, 45.
15. Gallup, 44–45.
16. Gallup, 45.
17. Australian War Memorial (hereafter AWM) 54 883/2/297 Part 1.
18. Churchill, *His Complete Speeches*, 6634.
19. AWM 54 883/2/297 Part 1.
20. These weekly reports were written by the censorship office to assess the morale of troops in combat based on their letters home. The reports assessed each national contingent separately, allowing me to assess each group's morale independently. The reports were intended to give commanders a sense of the attitude, needs, and concerns of the troops in the field.
21. AWM 54 883/2/97 Part 1.
22. AWM 54 883/2/97 Part 2.
23. AWM 54 883/2/97 Part 2.
24. AWM 54 883/2/97 Part 1.
25. AWM 54 883/2/97 Part 1. Italic added.
26. AWM 54 883/2/97 Part 1.
27. Atlantic Charter, August 14, 1941.
28. AWM 54 883/2/97 Part 1.
29. Bierman and Smith, *The Battle of Alamein*, 66.

30. Bierman and Smith, 64.
31. Maughan, *Australia in the War*, 50.
32. Bierman and Smith, 69.
33. Maughan, 51; Bierman and Smith, 71.
34. Maughan, 51.
35. Maughan, 51.
36. Playfair, *Volume II*, 28.
37. Playfair, 30; Bierman and Smith, 73.
38. Bierman and Smith, 73.
39. Bierman and Smith, 73.
40. Jackson, *Battle for North Africa*, 106.
41. Maughan, 105.
42. Maughan, 107.
43. Maughan, 107.
44. Bharucha, 154.
45. Bierman and Smith, 82.
46. Maughan, 130.
47. Maughan, 112.
48. Jackson, *Battle for North Africa*, 137.
49. Maughan, 137, 271; Bierman and Smith, 83.
50. Bierman and Smith, 80.
51. AWM 54 883/2/97 Part 1
52. Bierman and Smith, 81.
53. Bierman and Smith, 83.
54. Maughan, 271.
55. Maughan, 130.
56. Maughan, 349.
57. Pitt, *Crucible of War*, 319.
58. AWM 54 883/2/97 Part 1.
59. AWM 54 883/2/97 Part 2.
60. AWM 54 883/2/97 Part 2.
61. AWM 54 883/2/97 Part 2.
62. AWM 54 883/2/97 Part 2.
63. Neillands, *Eighth Army*, 64.
64. AWM 54 883/2/97 Part 1.
65. AWM 54 883/2/97 Part 1.
66. AWM 54 883/2/97 Part 1.
67. AWM 54 883/2/97 Part 1.
68. AWM 54 883/2/97 part 1.
69. Pitt, 355.
70. AWM 54 883/2/97 Part 2.
71. AWM 54 883/2/97 Part 1.
72. AWM 54 883/2/97 Part 1.
73. Playfair, *Vol. III*, 4.
74. Bierman and Smith, 105.
75. Neillands, *Eighth Army*, 75; Bierman and Smith, 107.
76. Playfair, *Volume II*, 63.

77. Neillands, 75.
78. Bierman and Smith, 108.
79. Neillands, 77.
80. Playfair, *Volume III*, 25.
81. Playfair, *Volume III*, 42.
82. Playfair, *Volume III*, 48.
83. Neillands, 80; Maughan, 439.
84. Maughan, 439.
85. AWM 54 883/2/97 Part 2.
86. AWM 54/883/2/97 Part 2.
87. AWM 54/883/2/97 Part 2.
88. Fennell, *Fighting the People's War*, 156–58.
89. AWM 54 883/2/97 Part 2.
90. AWM 54 883/2/97 Part 2.
91. AWM 54 883/2/97 Part 2.
92. AWM 54 883/2/97 Part 2.
93. Neillands, 102.
94. Bierman and Smith, 177.
95. AWM54 883/2/97 Part 2.
96. Churchill, *Speeches of Winston Churchill*, 6630.
97. Churchill, *Speeches of Winston Churchill*, 6630.
98. AWM 54 883/2/97 Part 2.
99. AWM 54 883/2/97 Part 2.
100. Neillands, 108.
101. Playfair, *Volume III*, 243.
102. Playfair, *Volume III*, 247.
103. Neillands, 113; Bharucha, 408.
104. Bierman and Smith, 184.
105. Bierman and Smith, 184.
106. Bierman and Smith, 177.
107. AWM 54 883/2/97 Part 2.
108. AWM 54 883/2/97 Part 2.
109. AWM 54 883/2/97 Part 2.
110. Fennell, *Fighting the People's War*, 181
111. Fennel, "Courage and Cowardice," 100.
112. Fennel, "Courage and Cowardice," 105.
113. French, "Discipline and the Death Penalty," 538.
114. French, "Discipline and the Death Penalty," 541.
115. French, "Discipline and the Death Penalty," 541.
116. Bierman and Smith, 184.
117. AWM 54 883/2/97 Part 2.
118. Grauer, "Why Do Soldiers Give Up?"
119. Barr, *Pendulum of War*, 85.
120. Barr, 85.
121. Barr, 159.
122. Bierman and Smith, 208.
123. Barr, 92.

124. Hasan, *Towards Freedom*, 316.
125. Bhattacharya, *Propaganda and Information*, 150.
126. Bhattacharya, 155.
127. Raghavan, *India's War*, 59.
128. Raghavan, 20.
129. Raghavan, 20.
130. Raghavan, 77.
131. Bayly and Harber, *Forgotten Armies*, 79.
132. Raghavan, 217.
133. Hasan, 261 and 289.
134. Hasan, 290.
135. Sarkar, *Modern India*, 371–72.
136. Khan, *India at War*, 58.
137. Hasan, 318.
138. Hasan, 260.
139. Raghavan, 252.
140. Hasan, 301.
141. Kaura, *Muslims and Indian Nationalism*, 138.
142. Hasan, 300.
143. Khan, 40–41.
144. Khan, 18.
145. Khan, 52.
146. Khan, 68; Raghavan, 47 and 257.
147. Cohen, *The Indian Army*, 95.
148. Khan, 72.
149. Khan, 141 and 143.
150. Trench, *The Indian Army and the King's Enemies*, 137.
151. Raghavan, 76.
152. Raghavan, 76.
153. British Library (hereafter BL) IOR L/P&J/12/654.
154. Cohen, 141.
155. Barkawi, *Soldiers of Empire*, 62.
156. Raghavan, 81.
157. Trench, 137.
158. Bharucha, 37.
159. Playfair, *Volume II*, 8.
160. Maughan, 77.
161. Maughan, 83; Playfair, *Volume II*, 28.
162. Bharucha, 148.
163. Playfair, *Volume II*, 28.
164. Jackson, *Battle for North Africa*, 106.
165. Jackson, *Battle for North Africa*, 108; Bharucha, 154.
166. Bharucha, 154.
167. Fennell, *People's War*, 144.
168. Bharucha, 166.
169. Maughan, 247.
170. AWM 54 883/2/97 Part 1, 94.

171. As quoted in Maughan, 335. Italic added.
172. Maughan, 280.
173. The National Archives (hereafter TNA) WO 169/3351.
174. TNA WO 169/3351.
175. TNA WO 169/3443.
176. TNA WO 208/673.
177. Raghavan, 63.
178. Raghavan, 81.
179. Raghavan, 74.
180. Bharucha, 210.
181. Bharucha, 228.
182. Bharucha, 234
183. Bharucha, 235.
184. Playfair, *Volume III*, 82.
185. Playfair, *Volume III*, 82.
186. Bierman and Smith, 116.
187. Bierman and Smith, 157.
188. Bierman and Smith, 166.
189. Playfair, *Volume III*, 223.
190. Bharucha, 370.
191. Playfair, *Volume III*, 234.
192. Playfair, *Volume III*, 234.
193. Barr, 85.
194. Barr, 85.
195. Barr, 78.
196. Barr, 79; Bharucha, 418.
197. Barr, 79.
198. Maughan, 549.
199. Maughan, 577; Barr, 134.
200. Barr, 146.
201. BL IOR L/PJ/12/654.
202. BL IOR L/PJ/12/654.
203. BL IOR L/PJ/12/654.
204. BL IOR L/PJ/12/654.
205. Younger, *Australia and the Australians*, 597.
206. Younger, 581.
207. Younger, 593.
208. As quoted in McLachlan, *Waiting for the Revolution*, 241.
209. Welsh, *Australia*, 423.
210. Younger, 598.
211. Younger, 599.
212. Cantril, *Public Opinion 1935–1946*, 1071.
213. Cantril, 1072.
214. AWM 54 883/2/97 Part 2.
215. As quoted in Johnston, *Fighting the Enemy*, 19.
216. Johnston, *Fighting the Enemy*, 21–22.
217. Johnston, *Fighting the Enemy*, 20.

218. Johnston, *Fighting the Enemy*, 32.
219. AWM 54 883/2/97 Part 1.
220. AWM 54 883/2/97 Part 1.
221. AWM 54 883/2/97 Part 1.
222. Maughan, 41.
223. Maughan, 67.
224. Maughan, 74.
225. Maughan, 75.
226. Maughan, 105.
227. Maughan, 107.
228. Jackson, *Battle for North Africa*, 107.
229. Maughan, 108; Bierman and Smith, 67.
230. Jackson, *Battle for North Africa*, 108.
231. Jackson, *Battle for North Africa*, 145.
232. Maughan, 150.
233. Maughan, 153.
234. AWM 54 883/2/97 Part 1.
235. Neillands, 48.
236. Playfair, *Volume III*, 22.
237. AWM 54 883/2/97 Part 1.
238. AWM 54 883/2/97 Part 1.
239. AWM 54 883/2/97 Part 1.
240. AWM 54 883/2/97 Part 1.
241. AWM 54 883/2/97 Part 1.
242. Maughan, 351. See also chapter 5.
243. Bierman and Smith, 82.
244. AWM 54 883/2/97 Part 1.
245. AWM 54 883/2/97 Part 1.
246. AWM SALT 175/3/2.
247. AWM SALT 175/3/2.
248. AWM 54 883/2/97 part 1.
249. AWM 54 833/2/97 Part 1.
250. AWM 54 883/2/97 Part 1. (Italic added).
251. AWM 54 883/2/97 Part 1.
252. AWM 54 883/2/97 Part 1.
253. AWM 54 883/2/97 Part 2.
254. AWM 54 883/2/97 Part 1.
255. AWM 54 883/2/97 Part 2.
256. AWM 54 883/2/97 Part 1.
257. AWM 54 883/2/97 Part 2.
258. AWM 54 883/2/97 Part 2.
259. AWM 54 883/2/97 Part 2.
260. AWM 54 883/2/97 Part 2.
261. AWM 54 883/2/97 Part 2.
262. AWM 54 883/2/97 Part 2.
263. AWM 54 883/2/97 Part 2.
264. Playfair, *Volume III*, 339.

265. Tibbitts, "Australians in the First Battle of El Alamein," 8.
266. Playfair, 339.
267. AWM 54 883/2/97 Part 2.
268. AWM 54 883/2/97 Part 2.
269. Playfair, *Volume III*, 346.
270. Barr, 109.
271. Tibbitts, 15.
272. Tibbitts, 18.
273. Tibbitts, 12.
274. Maughan, 589.

CHAPTER 4

Malaya

The Japanese invaded Malaya on December 8, 1941. In this chapter I consider the combat from that date until January 22, 1942, when the imperial forces evacuated the Malayan Peninsula to Singapore Island. I do not consider the limited engagements that occurred on Singapore Island from January 22 until the garrison surrendered on February 14, as the concentration of forces on the island make it very difficult to distinguish clearly between the different national units in specific engagements.

I rely on secondary sources and unit histories and diaries that were reconstructed after the battle for this chapter. Because of the speed of the Japanese invasion and the eventual surrender of all remaining soldiers, including headquarters, many of the original records were lost. Many of the unit histories I examine here were reconstructed by officers during their time as prisoners of war.[1] I was also unable to find censorship reports that covered this region in the archives. There were censorship reports for North Africa and Italy, as well as domestic reports, but I did not locate any for Malaya, Singapore, or Hong Kong. This means I have less access to soldiers' discussions of their circumstances for this case than for those in chapters 3 and 5.

The chapter concludes that British soldiers fought with adequate levels of will to fight, Indian soldiers with poor to adequate levels, and Australian soldiers with high will to fight. This variation is explained by the fact that while Australian soldiers saw this fight as a clear battle to defend the empire and Australia from tyranny by non-white peoples, British and Indian soldiers were less convinced that the battle was relevant to them. British soldiers were ambivalent toward the empire and were more concerned about defending Britain from Germany, while Indian soldiers were unconvinced that they should fight one empire to defend their subjugation under another.

BACKGROUND

The war in the Pacific began in December 1941, two and a half years into the war with Germany and Italy. Japan's expansionism had been evident for some time, forcing imperial officials to plan for war on two fronts. World War I left the Japanese and Americans as the most powerful forces in the Pacific, and Britain concluded that any future threat would come from one of these two states.[2] In order to defend its imperial holdings in the Pacific, the British decided to build a naval fortress at which to station a large fleet to deter any aggression against its Pacific possessions.[3] Construction of the fortress began in Singapore in 1919. Although the Singapore strategy, as this plan was known, became the only plan for Pacific defense, the resources available never matched the demands.

As tensions rose in Europe, Japan accelerated its expansion in the Pacific. That expansion posed a direct threat to Australia, India, and other British colonies in the region. Though the plan to defend against Japan focused on Singapore Island, a major naval base, the British Navy was concentrated in the Atlantic and Mediterranean to defend against Germany, making a naval defense strategy unworkable in the Pacific. Australia pressed for a fleet to be sent to Singapore during peacetime, but the British insisted that priority lay in the home waters and that the fleet would go as soon as it was necessary to defend the fortress.[4] After war broke out with Germany in 1939, Australian leaders again sought and received reassurance that should Japan attack British interests in the Pacific, a fleet would move to Singapore.[5] It was on this reassurance that the Australians agreed to send troops to North Africa in 1940.

The Australian government continued to lobby for greater preparation of Singapore through either naval or air reinforcements to no avail (though their own soldiers were much more interested in going to the Mediterranean to fight the Germans than in garrisoning Malaya).[6] Churchill's belief in the British government's own "Fortress Singapore" propaganda, as well as his focus on defeating Germany caused Malaya to be severely underequipped when the Japanese attacked on December 8, 1941.

The strategy for Singapore's defense was such that there was very little hope of success, given the realities on the ground. From the chiefs of staff all the way down to the division commanders on the ground, plans for defense were based on planned strength versus actual numbers, hoped-for warning from the Japanese, and an air force that simply did not exist.

Two more specific strategies were developed to hold Malaya until those hoped-for reinforcements arrived. First, Operation Matador was intended to hold the Japanese north of the Malayan border in Thailand. The Japanese had been increasing their pressure on the neutral Thai government, and the British military leaders assumed (rightly) that any Japanese invasion force would include a move through Thailand into northern Malaya. The British intended Operation Matador to counter this threat. The plan was to send troops north into Thailand as soon as they received notice of Japanese movements, where they would meet the Japanese and delay their approach into Malaya.[7] A good plan on paper, the politics of the situation never matched the military realities. Because Churchill did not believe the Japanese could invade through the jungle, he refused to increase troop numbers to guard against that possibility. Additionally, the assigned troops needed at least sixty hours to reach the necessary positions. The plan thus depended on the Japanese giving nearly three days' warning before the invasion.[8] Despite this unlikely assumption, Matador remained the main plan for Malaya's defense.

The second strategy was an air defense of the peninsula with the army acting in a supporting role by defending the airfields.[9] This plan required some 336 first-class airplanes.[10] By December 1941, a number of new airbases had been built to accommodate those planes, but according to one engineer, "there were nearly as many landing grounds as first class bombers."[11] British Air power was nearly nonexistent in the battle for Malaya.

Despite the clear political and military challenges to both of these plans, the imperial forces carried out troop dispositions and defense building according to both. A unit named Krohcol was assembled and sent north to await orders for Matador, while the rest of the British and Indian forces were dispersed to defend the three major airfields in the north.[12] The Australian troops were held further south to defend the western side of the peninsula from possible Japanese landings. According to the battle's official historian, once General Brooke-Popham and the chiefs of staff decided on these dispositions, there was little chance of success. The guiding strategy for defending Malaya depended on airplanes that did not exist and advance warning of Japanese intentions.[13]

By the time the Japanese invaded on December 8, 1941, there was a force of 80,000 British, Indian, and Australian troops in Malaya.[14] The British government began sending reinforcements in 1939, and troops of all three nationalities arrived throughout 1940 and 1941. On December 8, 1941, the following units were present: the 11th British Indian Army Division, consisting of nine Indian brigades and three British brigades, and the 8th Australian Infantry Division, consisting of four brigades. Despite their large numbers, they were not well equipped for war in the jungle. The troops were also short of artillery and antiaircraft units, most of which were in the Mediterranean.[15] There were some armored vehicles but few roads solid enough on which to drive them.

The soldiers stationed in Malaya arrived in various states of training. Most had been intended for the Middle East but diverted to appease the Australian government worried about the Japanese threat.[16] As a result, few arrived in the country with any knowledge of jungle warfare; in fact, few in the imperial forces as a whole had any real knowledge of jungle warfare until after the initial battles with Japan in Malaya.[17]

Unlike the North African campaign, there was no central training center for troops before they went to the front line. There was not yet any front line in Malaya and no jungle warfare doctrine in which to train the troops. As a result, training typically took place at the brigade level, and there was significant variation in the type and amount of training the various units in Malaya underwent.[18] British and Indian units tended to follow the pattern of garrisons in India—formal drilling in the morning and evening, with the hot afternoons reserved for recreation and rest.

Once soldiers arrived in Malaya, it was up to individual commanders to determine how best to train their men for battle in the jungle, and those commanders were far from certain themselves. The information the army did provide suggested that the Japanese would not be jungle trained either, reacted badly to surprise, and lacked initiative—racist information not likely to encourage hard training in uncomfortable conditions.[19]

There were three formations that did attempt to train for jungle warfare: the 22nd and 27th Australian Brigades and the 12th Indian Brigade.[20] Commanders in these units drilled on the need for speed, individual roles on engagement, and even just growing accustomed to the density of the jungle.[21] The 12th Indian Brigade, having been stationed in Malaya since 1939, had developed their system of training through several years of trial and error, mostly through the efforts of Lt. Col. Ian Stewart.[22] The more recently arrived Australian brigades drew on Stewart's experience as well as the more accurate information about the Japanese provided by the Australian government.[23] The training accorded to these three brigades would stand them in good stead when they met the Japanese in battle. However, as we will see below, level of training did not predict the will of a unit in battle.

BRITISH IN COMBAT

Perception of the War Goals

The Japanese posed a threat to Britain's Asian colonies that provided important resources for the ongoing war with Germany. Malaya in particular provided a significant amount of rubber and tin, which were central to military manufacturing.[24] Additionally, India provided a huge amount of Britain's military manpower and financing: India's military budget (paid by the Indian people and not London) was six times Australia's.[25] By 1941 the Indian Army had grown to 900,000 soldiers, many of whom were stationed outside of India in Burma, Malaya, and Hong Kong, and were actively fighting throughout Africa.[26] Yet despite relying on its Asian colonies for manpower, finances, and raw materials, these colonies were the subject of increasing contention at home.

British leaders downplayed both the likelihood of war with Japan and the difficulty that such a war might present. When Churchill discussed Japan in his public speeches, it was with reference to the Japanese tension with the United States and the British need to support America in that potential conflict.[27] Even after Japan attacked and the battle for Malaya was ongoing, in a speech to the US Congress on December 26, 1941 (a speech broadcast on the BBC throughout the world) he spoke primarily about Germany. When he mentioned Japan, it was to point out that defeating Germany was more important than preparing for defense against Japan: "If you ask me—as they have a right to ask me in England—why is it that you have not got ample equipment of modern aircraft and Army weapons of all kinds in Malaya and in the East Indies, I can only point to the victories General Auchinleck has gained in the Libya campaign. Had we diverted and dispersed our gradually growing resources between Libya and Malaya, we should have been found wanting in both theaters."[28] The movements of the Japanese on the other side of the world were less important than the need to defeat the Nazi threat.

The Labour Party also downplayed the role of defending the colonies in Britain's war goals. In 1940 the party published a short book titled *Why We Fight: Labour's Case*. In it, the party walks a fine line of supporting self-government and dominion status for India and other colonies while also arguing that Britain had encouraged democracy through its empire.[29] Referencing only Germany (indicating that even in 1940 Japan was far from the British mind), Labour argued that though the British colonies should be given self-government (at some unspecified time in the future), British rule was still preferable to what would certainly be barbaric German rule.[30] This became the British argument regarding Japan as well. Although an expansive empire helped to define British identity, when faced with a war to defend it, leaders had difficulty justifying its continued existence and expending British lives to keep it (though most of the soldiers posted to the colonies were Indian or other colonial soldiers by this point in the war).

Public opinion regarding the colonies was mixed. A Gallup poll conducted in November 1939 asked: "Should India's demand for self-government be granted during the war or should it wait until after the war?" Only 26 percent wanted to grant independence during the war, but an additional 51 percent wanted to grant independence after the war was over.[31] By January 1942, at the height of the fighting in Malaya, that had shifted to 31 percent during the war and 41 percent after the war.[32] The next month, February 1942, responders were asked, "Which country is the greater threat to the future of the British Empire, Germany or Japan?" Forty-seven percent responded with Germany and 31 percent with Japan.[33] The British public were not fully committed to India's place in the empire and certainly (understandably) saw Germany as a much greater threat to that empire even as Japan was the opponent actively dismantling it. Colonies such as Burma and Malaya were even more tangential to British self-identity than India; pollsters did not even ask the public about their status.

Soldiers in Malaya saw clearly the undemocratic treatment that colonial subjects there endured. The British actively repressed political expression among colonial peoples. British soldiers were used to break strikes on rubber plantations, an order Australian commanders refused.[34] Malayans, Chinese, and Indians were all barred from any high-ranking civil service position, no matter their education or qualification. There was no social interaction between Malayans and white colonists. British citizens (all white) lived in luxury and gave little consideration to developing their subjects' economy and infrastructure.[35]

Active repression of political activity also resulted in openings for Japanese espionage and organization among the national population. Throughout the 1920s and 1930s Japan's growing power attracted attention from nationalist leaders across Asia. Numerous Malayan and Indian leaders began studying Japan as a model, and some traveled to Tokyo and made contacts among Japanese officials. As Japan prepared for war these individuals were organized throughout the British colonies and did provide some aid to the Japanese when they invaded the territories.[36] The combination of racial prejudice, awareness of the empire's repressive policies, and evidence of Japanese infiltration made British soldiers extremely suspicious of the population they were supposed to be protecting. Numerous Malayan civilians were shot for suspected espionage and sabotage; some may actually have been guilty, but many were simply in the wrong place at the wrong time.[37]

Finally, the soldiers who fought the war against Japan had spent the last few years hoping to see action against the Germans. Germany posed the main concern to the British nation and was thought to be the more challenging and worthy opponent. One soldier described this pre-war attitude: "Even after Pearl Harbor we thought the Japs might be a bit Gilbert and Sullivan, sort of Oriental Italians and definitely not up to our standards. We wanted to biff the Hun."[38] The Japanese were not considered a real threat or even a worthy opponent. Even the reference here to Pearl Harbor as opposed to the attacks on the British-held territories suggests a disconnect from the empire.

Ambivalence about the empire and a lack of perceived threat combined to make British soldiers unenthusiastic about their fight against the Japanese. After the battle of Singapore ended in Britain's worst-ever military defeat, Lt. Col. Ian Stewart gave his own assessment of why the British lost: "Few in Malaya were on fire... Internal propaganda among the troops, as elsewhere, had received no official attention, and to interest the men in the war and to inspire them was left to unit commanders, who were given practically no facilities to assist them. In consequence there was virtually no emotional enthusiasm to carry the exceptionally high degree of physical exhaustion."[39] British soldiers were unclear on why they were fighting the Japanese so far from home and were not "on fire" for their task. Stewart suggests that the government put little effort into creating such a fire and that it showed in their battlefield performance.[40]

The Campaign

This section demonstrates that British troops showed adequate levels of combat motivation, occasionally dipping into poor levels. They were consistently disciplined but morale could falter. They did not consistently show initiative in battle, the third category of will. Both national identity theory and threat are confirmed, but threat is determined by national identity. Malaya was integral to the war effort from the perspective of natural resources but did not factor into British perceptions of their own identity and thus the threat to their nation.

The first major battle (December 9–12) that British units participated in was a disastrous defensive battle in Jitra in Malaya's north. Here soldiers from the 11th Indian Infantry Division—including the British 1st Leicester and 2nd East Surrey—attempted to prevent the Japanese from crossing the Sungei Bata River. These two units had made particular efforts to maintain high standards of dress and mess rules during their garrison in Malaya, though battle would quickly relieve them of their tidiness.[41] The Leicesters had been on standby for the Matador invasion into Thailand, but they were moved into defensive positions at Jitra, and their Division historian writes that: "Worst of all it lowered morale. The sudden change from an offensive to a defensive role before battle was joined made the troops wonder. The Leicesters . . . had spent the morning dashing out of and into their trains as Japanese Aeroplanes came and went, [and] murmured: 'Hell! Back to the effing mud hole.'"[42]

Once it began, the battle was particularly intense, with the Leicesters giving ground and then gaining it back. Even the clerks at headquarters took part in driving the Japanese back.[43] Despite their efforts, the Jitra position was not a good one as defenses were only partially constructed and the Leicesters were dug in with the river to their back.[44] Eventually, as panic spread among the troops in the rear and commanders issued orders to withdraw, the Leicesters' commanding officer was reduced to swimming back and forth across the river to maintain contact with headquarters.[45] The bridge having been blown prematurely, the Leicesters had to make their own way back to the British lines once ordered to withdraw; many were captured on the way.[46] Demonstrating discipline, those who escaped the Japanese eventually reassembled farther south at Gurun.

Meanwhile, those units that had been sent north as part of the Matador plan (see above) were fighting a rearguard action south, down the main road from Kroh toward Ipoh. Because Matador was ordered too late to prevent the Japanese from gaining a launching ground in Thailand, Krohcol (the unit assembled to meet the Japanese in Thailand) and the 12th Indian Infantry Brigade that reinforced them on December 14 sought to slow the Japanese move south while moving to regroup with the rest of the British forces in northern Malaya. As one of the best-trained units in Malaya, 12th Brigade as a whole held up well under the pressure of Japanese infiltration and encirclement tactics. While both of the division's Indian battalions performed well (and will be discussed in further detail below), the British 2nd Argyll and Sutherland Highlanders were "the stars of the rearguard fighting."[47]

The Argylls were the embodiment of British esprit de corps in Malaya. With a storied history dating back to the Crimean Wars, the Argylls were a regiment proud of their history and traditions. They even carried their regimental silver with them into battle.[48] Eventually the silver was hidden and Lt. Col. Stewart had to write to the Argylls' commanding officer at home apologizing for dishonoring the unit by its loss.⁴⁹ This unit clearly had a strong sense of esprit de corps, suggesting that they should have high will to fight in Malaya.

The combination of a strong sense of esprit de corps, long service together, and intense training helped the Argylls to achieve higher will to fight than other British units. In one instance during the Kroh Road battle, a small group of thirty-five Argylls, commanded by the battalion's boxing coach Company Sergeant Major Archie McDine, held up a detachment of Japanese soldiers ten times their number.⁵⁰ The Argylls demonstrated strong discipline under very difficult circumstances.

But despite their training and esprit de corps, even the Argylls were flustered by the unexpected. When the Japanese brought tanks into the battle at Kampar (December 28), the Argylls broke up and fled the battle.⁵¹ Discipline broke down when faced with surprise, although it was regained quickly and the unit reassembled and continued to fight their way southward. However, they were not able to adjust their tactics when those that they had been trained for failed against the enemy. The Argylls showed discipline but not initiative.

These two forces—Krohcol (of which the Argylls were a part) and 1st Leicesters and 2nd East Surreys who had survived the Jitra battle—met at Kampar, intent on holding the position there so as to protect the Kuantan airfield (December 30).⁵² At this point, due to losses at Jitra, two new battalions were formed out of the remnants of the 15th Brigade's former battalions. The Jats and the 1/8th Punjabis (along with Indian soldiers of numerous ethnicities who had been separated from their own units) became the "Jat/Punjab Battalion," and the 1st Leicester and 2nd East Surrey Battalions became the "British Battalion."⁵³ Additionally, Lt. Col. Stewart, the Argylls' commander, was promoted to command all of the 12th Brigade, leaving the Argylls in the care of his junior officers.

Despite having been in action for weeks with little rest and having had major changes in command and organization, each of these British units was able to delay the Japanese at Kampar. The British Battalion was encouraged to form a new sense of esprit de corps, one that included both the Leicesters and the Surreys. According to one officer's account of the 11th Division's battles in Malaya:

> In this its first battle since its birth, the British Battalion lived up to the finest traditions of the two Regiments whose men had been brought together by disaster to form it—the men of the Leicesters and East Surreys. In the short time between the Battalion's organization and its first battle, Lt. Colonel Morrison had permeated it

with an esprit-de-corps second to none; in this great achievement he owed much to the assistance he was given by Captain Wallace and Regimental Sergeant Major Meredith, both of the East Surreys... its spirit may perhaps be discerned from the answer which greeted a questioner who asked a man whether he belonged to the Leicesters or the East Surreys: 'Neither', was the man's reply, 'I belong to the British Battalion.'[54]

This assessment suggests that esprit de corps or secondary-unit cohesion can be built in a relatively short amount of time among soldiers who are in combat. The British Battalion's history suggests that cohesion may be the result of experiencing battle together rather than a necessary preexisting source of motivation.

Beginning on January 1, 1942, the Japanese repeatedly attacked the British Battalion. According to their own war diaries, the Japanese had not expected the level of resistance they met.[55] Demonstrating discipline, the British Battalion stood firm in the face of superior artillery and even airpower (of which the British were seeing very little from their own side), and when the unit gave ground it did so under orders.[56] Even after their commanding officers were killed, junior officers took over smoothly and discipline was maintained.[57] This newly formed unit demonstrated morale and discipline in battle despite being an amalgamated unit and experiencing turnover among their officers.

The Argylls also continued to demonstrate adequate will, even after their longtime commander, Lt. Colonel Stewart, was promoted. They were not dependent on a particular leader for their cohesion. Morale remained high, perhaps encouraged after having found a supply of whiskey in Kampar with which to ring in the New Year before going into battle the next morning.[58] The Argylls continued to stand firm against stronger Japanese forces, even after their panic in the face of Japanese tanks a week earlier. Changes in leadership and near constant retreating did not lower their morale or undermine discipline in the battle at Kampar, where they held off the Japanese as the British Battalion withdrew south to the Slim River.

The withdrawal was puzzling for all troops involved, as the British Battalion had been successful in stopping the Japanese on the Trunk Road (the main north-south road on the peninsula). General Paris, commanding the division, feared that the Japanese would encircle the lines and cut off his supply route. He requested and received permission to withdraw to the Slim River, a supposedly prepared defensive position. Despite the confusion, discipline remained firm. The Argylls covered the retreat and then fought a rearguard action themselves to regroup at Slim River.⁵⁹ The British Battalion continued south, to fight again later at Batu Pahat.

The battle at Slim River (January 4–7 1942) was an unmitigated disaster for the two brigades ordered to defend the position—the 12th and 28th. Both had been in battle or marching constantly for three weeks, communications were unreliable at best, and the British generals directing the battle continued to discount the Japanese ability to use unpaved jungle roads to outflank and surround fixed defensive positions. At the end of the battle the 12th Brigade was no longer a fighting force, and "28th Brigade was standing; that was all."⁶⁰

The only British unit that fought at Slim River was the Argyll and Sutherland Highlanders.⁶¹ Having fought for nearly three solid weeks, all of 12th Brigade's men were exhausted and morale was beginning to deteriorate.⁶² Though tired from having been digging trenches by night and being bombed by day, the Argylls still demonstrated good attitude toward the fight and willingness to engage the enemy.⁶³

The Argylls were stationed in depth on the road and the railroad line leading south near the village of Trolak. The Japanese smashed through the Hyderabad and Punjabi units stationed just north of the Argylls and continued on to the Argylls' position across the road and rail line. On short notice they were ordered to construct roadblocks to prevent the passage of the Japanese tanks and hastily did so.⁶⁴ However, with inadequate armor and anti-tank weaponry, the tanks quickly smashed through the roadblock and continued south over the undemolished bridge. According to the official historian, "There was little that the Highlanders could do but watch helplessly as enemy tanks drove southwards."⁶⁵ Given the lack of communication and insufficient weaponry, it is unsurprising that the British troops felt helpless in the face of Japan's superior armor. However, they had known for several weeks now about their lack of success against Japan's tanks yet persisted in constructing roadblocks and following techniques proven ineffective.⁶⁶ The 11th Division historian notes, "The tanks were the cause of a great demoralization for which there was every excuse."⁶⁷ As we will see in the discussion about Australian will to fight below, with some initiative it was possible to defend more effectively against tanks. Even this most well-trained of British units lacked the initiative to try new tactics when the old failed.

After the Japanese tanks passed through their defenses, the Argylls were surrounded. They were ordered to retreat through the jungle, and small groups attempted to make their way back to the lines south of the Slim River. After the battle only 430 officers and men remained of the 12th Brigade.⁶⁸ The others had been captured or killed in the weeks of traveling through the jungle after the battle, though a few survived and joined guerrillas fighting behind enemy lines for the remainder of the war.[69]

The British units that fought in Malaya had adequate levels of combat motivation. Their morale remained high, even after considerable time in battle, and they maintained good discipline in the face of confusion, changes in leadership, and near constant retreating. However, they showed little initiative in battle. Though they did not avoid it, neither did they seek out engagement with the enemy through patrolling. British troops continued to use the same tactics that failed over and over again, failing to adjust to the Japanese tactics.

Lt. Colonel Ian Stewart wrote an assessment of British and Indian performance after Malaya fell. He wrote: "Skill by itself will not obtain to the topmost limits of practical efficiency; it will only do so when accompanied by ardent enthusiasm. Can we say that we are emotionally on fire to the extent of the youth of Nazi Germany and of Japan? Possibly when under the close stimulus of fighting near our homes. But, in the far east, all ranks were emotionally cold against an opponent who was ardent."[70] Stewart, who led one of the best-trained units in Malaya, knew the importance of jungle skills. Yet in his opinion the problems in Malaya were problems of a lack of emotional commitment to the cause in addition to problems of training and material. Without a perception of a link between their national identity and the fight against Japan, British soldiers demonstrated morale and discipline but not initiative.

INDIANS IN COMBAT

Perceptions of the War Goals

India did not face direct military threat during World War II until the Japanese attack on December 8. As the Japanese racked up successes in the Pacific, taking Hong Kong and steadily marching toward Singapore, the threat of invasion loomed large over India. By mid-February, when Singapore surrendered, the military was planning for the possibility of a major invasion of India.[71] Soldiers in Malaya knew that the defense of the empire—including India—depended on the defense of Singapore.

As noted in chapter 3, most Indian political leaders opposed German and Japanese imperialism. But while the Muslim League and some Hindu nationalists supported the British war in an effort to gain British support for their independence plans, Congress opposed the war.[72] There were, however, some nationalists who saw Japan as an Asian power that had succeeded in challenging the West and that might aid India's quest for Independence.[73] Throughout the 1930s Tokyo hosted several nationalist leaders from Britain's Asian colonies, including Indian nationalists.[74] The Japanese saw them as potential fifth columnists, though this only came to fruition in a few cases, including in Malaya. One well-known Indian the Japanese recruited was Subhas Chandra Bose.

Bose was a high-ranking official in the Indian National Congress and one of the most vocal opponents of the British war effort.[75] In 1940 he escaped to Moscow from house arrest in India and traveled on to Germany, where he worked with the Nazis to form a unit called Hitler's Indian Legion, made up of a group of anti-colonial Indian prisoners of war captured in North Africa.[76] When the Nazis refused to try the unit in battle against the British, Bose returned to India and then went on to Japan. He had more success with Indian POWs captured by the Japanese, with whom he formed the Indian National Army (INA). Made up of captured soldiers (some of whom volunteered while others were coerced), the INA did fight against the British in Burma in 1943 and 1944 but was never particularly successful. Bose did, however, offer an alternative to support for the British Empire.

British war propaganda in India sought to discredit Congress. British officials sought to portray the party, with its civil disobedience movement and anti-British position, as appeasers opening India to fascism and potential attack.[77] Additionally, the British claimed that the war itself was morally good, that India would benefit from the industrial growth resulting from participating in the war effort, and that independence would come "soon after" victory: helping Britain to win the war would improve India's economy and speed the path to independence.[78]

Both arguments failed. Popular support for Congress continued throughout the war, and the British leaders' continued efforts to discredit the party only led to greater sympathy, even from its traditional competitors. British efforts to discredit Congress served only to drive various nationalist camps closer together.[79] Support for Congress was widespread in India, including in the princely states that were governed centrally by a prince or maharaja.[80] Many Indians followed Gandhi and Nehru in their distaste for fascism as well as their desire not to give the British more aid than necessary, though neither did they want to see India under Japanese control.[81]

When Indian soldiers arrived in Malaya, they experienced firsthand their status as second-class citizens. While Indianization of the officer corps resulted in new openness for Indian officers in India and North Africa, in Malaya they were subject to the same racial barriers as non-whites in Malaya and prevented from participating in British society.[82] The color bar was policed aggressively in Malaya and affected all soldiers stationed there.[83]

Additionally, they were witness to discrimination and mistreatment of Indian workers by both British and Malayan elites. Ethnic tensions among Indians, Malayans, and Chinese ran strong in Malaya, and Indian soldiers were appalled at the treatment of their fellows in this part of the empire. This discrimination served to drive Hindu and Muslim soldiers together, breaking through the continued social barriers between these two groups and uniting them as Indians.[84]

Indian national identity was divided by class and religion. Yet the soldiers in Malaya also witnessed clear discrimination against the local population and experienced more discrimination themselves than they were used to even in India. Indian national identities were not compatible with the goal of defending the empire and British power against Japan, and they rejected British claims that this was a war to defend or promote democracy against fascism. Even the direct threat to their homeland was not enough to lead them to support British over Japanese imperialism. They were caught between two unappealing options.

The Campaign

Indian soldiers showed poor to adequate combat motivation in Malaya. Some units, especially those who had served together with their officers in Malaya for some time, were adequate in battle. Other units had low levels of discipline and morale and showed no initiative in battle. More than either the British or Australian troops, Indian soldiers depended on experienced officers and a sense of esprit de corps for their motivation, and were likely to collapse in battle without them.

Indian units were among the first to meet the Japanese in Malaya. The 8th Indian Infantry Brigade was stationed around the Kota Bharu airfield with the goal of allowing the air force to operate from that base for as long as possible. The Japanese landed on the beaches near Kota Bharu early on the morning of December 8. The 8th Indian Infantry Brigade had some air support, though it had little effect on the Japanese landings. The 3/17th Dogras put up strong resistance, but the Japanese made good headway and reinforcements were not able to get through the swampy terrain quickly enough to counterattack.[85] Brigadier General Key called the 3/17th back from the beaches to defend the airfield more closely with the rest of the brigade.

At the airfield Indian units showed varied levels of will in a situation of much confusion. As Key brought his brigade back from the beaches to defend the airfield at close quarters, the sounds of the battle neared as well. The Royal Australian Air Force (RAAF) ground crew at the airfield thought that Japanese attack was imminent and obtained permission from their headquarters to abandon the airfield. Key was not told of their withdrawal through his own lines, and they caused a great deal of confusion among his soldiers. Most units displayed discipline by remaining in position and continuing the fight, but at least one unit fled the battle when they saw the airfield staff leaving. The 1st Hyderabad State Infantry, an Indian States Force unit "donated" to the Indian Army by the Hyderabad state prince, left their lines and fled south. Their commanding officer and adjutant were killed, with some suggesting that the officers were victims of their own unit.[86] The 2/12th Frontier Force, positioned nearby, waited until they received orders from the commanding officer and withdrew south with good discipline.[87] Both units witnessed the RAAF forces abandoning the airfield they were supposed to be protecting, but the Hyderabads fled while the Frontier Force remained to fight. A major difference between these two units was that the Frontier Force had a large contingent of British officers and a long tradition of esprit de corps that the Hyderabads lacked.[88] Additionally, the Hyderabad region of India had been heavily influenced by the Indian National Congress and Gandhi's nationalism, and soldiers from the region were known to have strong nationalist leanings.[89]

The next major defensive stand took place at Jitra (December 9–12), where some fixed defenses had already been prepared. When Matador was called off, the 11th Indian Infantry Division moved to Jitra to establish a defensive line and discovered much work still to be done on the positions; water had to be cleared from the trenches, and anti-tank mines and barbed wire needed to be laid.[90] Major-General Murray-Lyon, commanding the area, realized he would need at least two days for that work, and so on December 10 sent the 1/14th Punjabis and 2/1st Gurkhas north to Asun (December 9–11), where a narrow road and creeks offered a good delaying position.[91] The 1/14th Punjabis were positioned several miles north of the creek behind which the Gurkhas had established their position.[92] After their first battle with the advancing Japanese, the 1/14th Punjabis were ordered to disengage and withdraw back to the Gurkha's position at the stream. The plan was to cross south of the stream and then blow up the causeway, forming a tank barrier. However, before the Punjabis could complete their withdrawal, Murray-Lyon ordered them back to their position and to delay the

Japanese as far north as possible. Murray-Lyon was trying for more time to prepare the Jitra positions, but the 1/14th Punjabis were frustrated by the constantly changing plans.[93]

The Japanese attacked the Punjabi position that afternoon in the midst of an intense rainstorm.[94] Having been obscured by the rain, the tanks that appeared on the road shocked the Punjabis. Without attempting to organize their defenses or anti-tank weapons, the Punjabis scattered and fled south.[95] Despite the efforts of their British officer and physical attacks by their Indian CO Captain Mohan Singh, the Punjabis dispersed into the jungle, many of them fleeing through the 2/1st Gurkhas behind them.[96] The Gurkhas put up some resistance, with one officer managing to hit and stop one tank with an anti-tank rifle.[97] However, the 2/1st Gurkhas withdrew without orders, and "for the time being this battalion was as good as useless."[98] Out of 600 in the battalion, 30 were killed and 200 held together by the battalion's British commanding officer and returned to the Jitra line; the rest fled into the jungle to trickle back south over the next few weeks.[99] As the official report mildly put it, "this first calamity affected the morale of the Indian troops of the Division."[100]

Neither of these two units had much will in battle. While most soldiers—British or Indian—were shocked the first time they encountered Japanese tanks in the jungle, the Punjabis' flight was chaotic and final. Their commanding officers were unable to reassemble the unit for further action.[101] While the Gurkhas officer was able to stop a tank with an anti-tank rifle, his men withdrew without orders. Both units displayed a lack of discipline and morale. Both units suffered from "milking"—the practice of taking experienced officers from established units and moving them to newly established units in the expanding Indian Army—and lacked experienced NCOs.[102]

The Japanese were already trying to use nationalist divisions between Indian soldiers and their British colonizers. Two of the 2/1st Gurkhas above who retreated from Asun were captured and then released. According to the official report:

> They said that they were taken under an escort to an [Japanese] H.Q. where there were Sikhs in civilian clothes wearing an armband. These Sikhs questioned them. A Japanese officer then gave them a paper on which was written words to this effect: 'We do not wish to fight Gurkhas but only the English who have no business in Malaya." They were told to go back to their unit and tell their comrades to cease fighting for the English.[103]

There are conflicting reports as to the efficacy of such propaganda efforts among Indian troops, but it is clear that the Japanese began their recruitment for the INA early on in the battle and had some aid from local Indians as well as Indian Army POWs such as Major Mohan Singh, who would go on to lead the INA.[104]

The battle at Jitra was another where Indian units showed varying degrees of will to fight. Some units performed well in battle, while others fled even the sound of battle and had to be turned back at gunpoint. Despite Major General Murray-Lyon's attempt to delay the Japanese at Asun, the Jitra position was still a poor one for defense by the time the Japanese attacked. The line stretched across twelve miles of hill country and included both the trunk [main] road and the railway line.[105] On the right flank the 2/9th Jats were stationed alongside the Leicesters, and on the left were the 2nd East Surreys and 2/16th Punjabis.[106]

As discussed above, the British units at Jitra held up under fire and put up a good, though short-lived, defense. The Indian Army units present were not as consistent. On the right flank, the Jats came under fire and panicked, sending inaccurate reports about being overrun back to headquarters.[107] Brigadier Lay responded by sending the 1/8th Punjab Brigade and two companies of the 2/16th Punjab Brigade from the reserve to the right flank.[108] This movement proved unhelpful. The 1/8th Punjabis were in a state of extremely low morale. Maj. Gen. Murray-Lyon observed them marching toward the battle and stated, "I haven't seen that look on men's faces since March '18."[109] When the 2/16th Punjabis came under fire, one company scattered into the jungle and could not be reassembled.[110]

There was a great deal of confusion among the Indian units in Jitra. However, there were also examples of firm discipline and commitment to the battle when led by experienced officers.[111] The 2nd Jats put up a strong fight even as the Japanese overwhelmed their lines: "D company faced the enemy with bayonets fixed. The blast of whistles rent the air; the enemy charged. Few survivors emerged from the gallant stand of D company of the 2nd Jats under Captain Holden, who was himself last seen hurling grenades at the charging hordes."[112] In an army that trained both officers and enlisted that the officer-man relationship was key to performance, experienced officers were key to ensuring discipline within the unit.

Worse was seen when these units were withdrawn to the River Bata. Those that remained of the British units came back in dribs and drabs but in good order. The remnants of the Indian units returned in panic, which spread easily to the units already positioned at the river, making everyone jumpy.[113] When a Japanese motorcyclist made it through the lines and across the bridge over the river, Murray-Lyon ordered the bridge blown for fear that tanks would not be far behind. The sound of the exploding bridge was too much for those members of the 2/16th Punjabis that had made it back to the river. Though they had been resting in the shade, they stood and started to run down the road away from the blast. According to Murray-Lyon, "It was only by getting into my car and getting ahead of them that I managed to stop them and turn them back. In the case of the 2/16th Punjab companies I actually had to threaten some men with my pistol before they would stop."[114] Discipline was weak among these units, with the threat of force necessary to keep them in order. Though there were instances of Indian units standing firm in battle, they were generally those units that had experienced officers and had been together for some time, such as the 2nd Jats. Those units without such leadership, either due to experienced leaders being removed to new units or casualties, were liable to collapse when facing the enemy.

After Jitra, the retreat to Gurun was messy, and the remnants of the 15th Brigade that had fought at Jitra were of no use to the troops already positioned at Gurun. When they were inserted among the other units of 11th Division, they caused panic and refused orders:

> During the next two hours Brigadier Garrett's main preoccupation was the reception of the un-reconditioned stragglers who had been sent forward from the rear to rejoin their units. These stragglers, when ordered to debuss and go forward, did not do so. In Brigadier Garrett's own words, "they merely became a menace to our side by shooting wildly in the direction of the 6th Indian Brigade's position." Panic-stricken stragglers also came streaming back in the reverse direction from the front.[115]

Officers who tried to stop these soldiers from fleeing their positions met with no success and eventually gave up.[116]

Not all the Indian troops were in such an undisciplined state. The Gurkhas of the 28th Brigade, having made a fighting retreat from Jitra to Gurun, then aggressively fighting a battle at the Gurun River, were exhausted but in good spirits at the end of the battle. "Hopes of twenty-four hours' rest were exploded [by orders to move]. Brigadier Carpendale reported that his Gurkhas were extremely tired. Lt. General Heath looked at them. Most of them had settled down cheerfully and were eating their first regular meal for many days; some were already washing their clothes; a few were asleep. They were giving little evidence of their inevitable exhaustion, and undoubtedly there was nothing wrong with their morale."[117] The Gurkha unit still had its experienced officers and had served together in Malaya for some time, while the soldiers of the 15th Brigade were without their officers and were new to their units. Morale and discipline were closely linked to the tenure and experience of unit leaders.

After the battle at Gurun, the 15th and 6th Brigades were in pieces.[118] As noted earlier, the East Surreys and Leicesters were combined to create the British Battalion, and the 1/8th Punjabis and the Jats were consolidated into the Jat/Punjab Battalion.[119] Despite the best efforts of the British Indian Army High Command, the diversity of India's religions and ethnicities was present in this new battalion.[120] However, the new structure and a few days' rest at Kampar served these soldiers well. They were fed, received new uniforms and some weapons, and had time to sleep.[121] New officers took over and sought to engage the men's martial honor: "Prior to the assault the company [of the Jat/Punjab Battalion] was addressed by Captain John Graham, the battalion's second-in-command. Graham told them that the situation was critical. Though few would witness their actions, the honor of the regiment depended on their success."[122] Just in case they needed more than unit honor to sustain them in another engagement with the Japanese, every man was given two drams of rum.[123]

Once again, the Battle of Kampar (January 1–3) showed the varied levels of will of the Indian Army units. The Jat/Punjab Battalion at Kampar had higher will than its original parts at Jitra or Gurun. The rest and reorganization enabled them to better follow orders and stay organized in battle. According to the after-action report from the battle, "Indian [battalions] of the 6th and 15th [brigades] were all right. Sikhs fought very hard."[124] As they had been at Jitra, the 2/9th Gurkha battalion was aggressive, organized, and defended their position against the Japanese until they were ordered to withdraw for fear of an outflanking maneuver.[125] The well-trained Indian units of 12th Brigade also fought well, springing an ambush on the Japanese and covering the retreat away from Kampar.[126] But the 4/19th Hyderabads withdrew from battle in considerable disorder, causing fear and lowering morale of the units through which they passed.[127]

At the disaster at Slim River (January 4–7), most of the 11th Indian Division was decimated. The Indian units once again proved to have varying levels of will to fight. The 12th Brigade, including the Argyll and Sutherland Highlanders, 5/2nd Punjabis, and 4/19th Hyderabads were together in covering the Trolak area. All were suffering from having been in battle for three straight weeks with very little rest; the Hyderabads were only a shadow of their former strength. Even the 5/2nd Punjabis, who until this point had withstood battle well, began to lose their morale.[128]

Both the Hyderabads and the Punjabis successfully repelled the first wave of attackers on January 5 and 7. Yet, rather than take heart from their success, both units became even more nervous. The Punjabis commanding officer even threatened punishment to one of his Indian Viceroy Commissioned Officers for spreading despondency among the troops.[129] He then reported to Brigadier Stewart that his unit ought to be withdrawn as they were likely to break under another attack.[130] Unfortunately, there was no one to replace them. Though having fought for the same time in the same conditions, the British Argylls were in much better spirits. Despite exhaustion, their morale remained intact.[131]

On the evening of January 7, the Japanese, now reinforced by sea, attacked the Hyderabads and Punjabis again. This time the defense collapsed, and the troops retreated. Some Indian NCOs did engage Japanese tanks, but as individuals with only grenades, had no success.[132] Meanwhile, their units disintegrated. Once again, the 4/19th Hyderabads fled battle and spread panic to the troops through which they passed, in this case the Punjabis.[133] The Punjabis' officer was able to collect some of his own men and withdraw south, but a great number of the Hyderabads and Punjabis disappeared into the jungle.[134] The 2/1st Gurkhas had been pieced together again after Jitra and reinforced with experienced junior officers but had no anti-tank weapons. The Japanese tanks destroyed the unit. Of the 500 men in the 2/1st Gurkhas, their officer could find only 12 in the aftermath.[135] The Indian units at Slim River had been through a number of battles already—some had demonstrated morale and discipline, such as the Punjabis, while others had broken, such as the 2/1st Gurkhas. Their efforts in this battle demonstrated that compared to British units who had fought for the same amount of time, the Indian units had lower discipline.

While all troops were discouraged by the lack of air support compared to the Japanese air attacks they often experienced, the Indian soldiers took it especially hard and as a signal of their value to the British. The following exchange recorded in the unit history is indicative of the Indian soldiers' concerns:

B[ritish] O[fficer]: (Cheerfully) You shouldn't worry about not seeing our aeroplanes. They're doing their job far away where you can't see them.

G[hurka] O[fficer]: (drily) In England, perhaps?

BO: (Crossly) No, here in Malaya, on the roads to Singapore, on the enemy on those roads, on their aerodromes and ships.

GO: (quizzically) Of course, sahib. The Japanese are very bad soldiers. Their aeroplanes come and bomb us and let their troops see them.[136]

Indian soldiers saw the lack of material resources the British government was willing to send to Malaya as a sign of its commitment to India and to Indian soldiers fighting for the empire. There was little reason for these soldiers to be committed to such an empire.

As the retreat southward toward Singapore Island continued, some reinforcements were added to the depleted forces in Malaya, and the Australian troops joined in the fight, having been held back to provide defense of the western coastline. The exhausted 11th Division units were withdrawn through those lines to proceed to Singapore and rest before joining the defense on the island.

The 45th Indian Brigade arrived from India with only basic training, having expected to go to Egypt for final training and to join the desert war.[137] The officers were new and did not speak the languages of their Indian soldiers.[138] They were sent to defend the Muar River crossing with little support (January 15–16). When the Japanese attacked, they targeted the officers who were trying to rally the troops, leaving the soldiers with little direction.[139] They retreated in disarray only to be mocked as the "Galloping Garhwalis" by the Australians through whom they passed.[140] Despite being reinforced by the 2/29th Australian Battalion, the 45th was surrounded and eventually had to leave the wounded behind and disperse into the jungle to escape the Japanese. Many died or were captured, and the wounded, whom the 45th had expected the Japanese to treat as POWs, were murdered on capture after being tortured.[141]

The last battle Indian units participated in on the peninsula saw a brigade decimated and demonstrated the importance of officers in the Indian Army; without experienced leaders, units were easily broken. With little commitment to the battle and dependent on their officers for discipline, the 45th Brigade was no match for the Japanese.

Indian units demonstrated adequate motivation when they had a strong sense of esprit de corps and were led by experienced officers. When those officers were killed or promoted out of the unit, Indian units lost their discipline easily. Their morale also suffered more easily than did the British units who fought with them. And like their British colleagues, Indian units did not show initiative in battle, continuing to use the same tactics and not seeking engagement with the enemy.

AUSTRALIANS IN COMBAT

Perception of the War Goals

Australians saw Japan as a threat to the empire they were committed to, their racial identity, and possibly to their own homeland. The government drew on race and empire to rally the people and soldiers to the cause of fighting Japan. The military publication *SALT* (distributed to Australian soldiers) noted on December 15, 1941: "For obvious reasons, Australian eyes are fixed on what the rest of the world calls the Far East, but to us is the Near North."[142] Geography meant that Japan was the more immediate threat to Australia itself, and the Australian government had traded troops sent to North Africa for promises of defense via the Singapore strategy. Just as with India, it was not entirely clear whether Japan aimed to attack Australia proper. As *SALT* put it in the next issue, on 21 December 1941: "Japan's projected and partly achieved New Order in East Asia (or Co-Prosperity Bloc) has meant in effect Japanese domination of the whole of the Western Pacific from Siberia to Australia—and definitions have been doubtful enough to include us."[143] Even as the Japanese stacked up victories against the British in Malaya, the Australian government was not entirely sure that Australia was a future target. But the possibility existed.

There was clear concern that Japan posed a potential threat to the Australian homeland but also concern about the threat to Australia's sense of identity. The same troop magazine described the Japanese plans for a co-prosperity sphere in very clinical terms, noting that it was unclear whether the sphere would include Australia but would very clearly infringe on Australia's economic freedom as well as repress the populations that they did bring under their control. Managing to sound both racist and concerned for democracy and self-determination, *SALT* argued on October 13, 1941: "Japan today stands as the only industrial and relatively advanced country in a vast area populated mainly by over-crowded and unresisting peasants. In that situation, she [Japan] aims, as Germany does in Europe, to create a self-sufficient area, with herself as the central industrial state, and dependent agricultural countries grouped around her."[144]

Besides having a security interest in maintaining the British Empire in the Pacific, Australians also saw the empire—and their participation in it—as beneficial to the other inhabitants of the region. *SALT* portrayed Japan as interested only in extorting resources from the regions they controlled, not in the improvement of the people there. Just before the Japanese attack on

Malaya, on December 1, 1941, *SALT* unselfconsciously gave the following description of the democratic nature of British rule in Malaya:

> His [Raffles, Singapore's first British leader] ideas of administration were equally sound. Although Singapore's government was in the hands of British officials, Raffles allowed magistrates, chosen from the British community, to participate in the administration, while a Malay headman was appointed to assist in native affairs. In the courts he insisted that European, Malay and Chinese immigrants all be treated fairly. . . . Finally, Raffles founded the Malay college at Singapore, for he believed that, in educating the Malays, "we shall lay the foundation of our dominion on the firm basis of justice and mutual advantage, instead of the uncertain foothold of force and intrigue."[145]

According to the publishers of this military journal, Singapore had been brought under British control to promote democracy and the betterment of its population. The case for defending Malaya and Britain's other colonies in the "Near North" rested not just on defending Australia but also on defending the ideals of democracy and development that Australians believed the empire to promote. The government argued for the war not just in terms of Australia but in terms of the empire that made up an important part of Australia's national identity.

Having not formed a wartime unity government as the British had, the conservative government lost an election in October 1941, and the new prime minister, John Curtin, had only just taken over when the war with Japan began in December.[146] The change in government was likely related to a lack of public confidence in the way the war was being run. In October 1941, a poll found that only 35 percent of Australians were satisfied with the way the government was running the war.[147] More evidence that the general population was suspicious of the government was their rejection of government propaganda. In March and April 1942, just after the fall of Malaya, the government launched an intense hate campaign against the Japanese. The *Sydney Morning Herald* editorialized that Australians did not need "a torrent of cheap abuse and futile efforts in emulation of . . . Goebbels."[148] After that, the government moved away from attempts to demonize the Japanese through propaganda. Government propaganda was rejected even as Australian soldiers openly expressed racist views, suggesting that soldiers' attitudes toward the Japanese were not the product of a recent spate of propaganda. Rather, they were long-held views finding apparent confirmation in current events.

Soldiers in Malaya saw the fight against Japan as very compatible with their national identity. According to Lachlan Grant, Australian soldiers writing home often framed their description of Malaya and Singapore in terms of Kipling, glorifying British imperial rule and the exoticism of their location.[149] Australian soldiers were hyperaware of racial differences, emphasizing the variation in skin tones in a region populated by several different ethnicities.[150] Class distinctions that were maintained by the British colonials to the disadvantage of the Australian soldiers made it even more necessary for Australians to assert their whiteness and thus their place in the racial order of the empire.[151]

Just as the poverty and repression visible in Malaya made British soldiers question the benefits of empire, so too were Australian soldiers disturbed by the status of the local population. Soldiers commented on the poor living conditions of all but the British elites, poor pay and labor conditions, and the seeming "un-British" class exclusions of the elite institutions.[152] The Australians may have had unrealistic expectations of equality throughout the empire, but most were willing to assign more of the blame to the local population than to imperial policies. While they were disturbed by the surprisingly poor economic and political conditions in Malaya, soldiers commented consistently on the squalor and disease in the more "authentic" parts of the country they toured, living conditions that appear to have reinforced racial stereotypes.[153] Government propaganda that emphasized the benefits of empire for the region meshed with racism to blame the local population for any political and economic problems the soldiers observed.

Racism carried over to soldiers' views of their Japanese opponents as well. Scholars have noted a distinction between Australian soldiers' attitudes toward their German and Japanese enemies: "Australian soldiers commonly felt hatred for the Japanese but compassion—or at least respect—for European enemies, and they were aware of that contrast both during and after the war."[154] This attitude toward the Japanese can be traced back at least as far as the turn of the century and stemmed from Australian identity as white in a region of the world where they were a minority, as discussed in chapter 2. The Japanese were particularly feared and hated, but Black aboriginals, Pacific Islanders, and Chinese all faced discrimination from a government and population determined to protect "white Australia."

By the time the Japanese attacked Malaya, Australian soldiers had been an integral part of the British Empire's war effort for over two years. They fought in North Africa and in Greece, volunteering to defend the empire from distant threats. Australia was closely tied to Britain and dedicated to aiding it in the war against Germany and Italy.[155] Soldiers who volunteered to fight abroad looked forward to fighting the impressive Germans in the Middle East. When the fight with Japan began, Australians were already dedicated to the cause of imperial defense. The war with Japan only made more obvious the relationship between imperial and Australian defense.

The war with Japan was compatible with Australia's national identity. First, it was the realization of long-standing geographic and racial fears. Australians saw themselves as an outpost of white civilization in the Pacific that was threatened by growing Japanese power. Second, Australians considered Japan's co-prosperity sphere to threaten the security and growth of democracy and "civilization" in Asia, just as Germany was threatening it in Europe. Thus race and geography, empire and democracy were all elements of Australian national identity compatible with the war against Japanese expansion in the Pacific. Malaya was the first place in which Australian soldiers had a chance to fight this new enemy.

The Campaign

Australian units had high levels of will in battle. They had high morale, were disciplined in battle, and exhibited initiative on the battlefield by trying out new tactics and actively seeking out engagement with the enemy.

The Australian Army in Malaya has acquired a bad reputation among historians. Civilians and naval personnel who escaped the island before its surrender reported that Australian soldiers were drunkenly carousing in Singapore, attacking civilians, and attempting to commandeer passage on departing boats. Reports of indiscipline led to rumors of cowardice and lack of will in battle. According to the editor of the *Straits Times*:

> It is with great reluctance that I pursue this question of the behaviour of some members of the A.I.F., but there is little likelihood that these notes will be helpful if they are not frank. There were desertions. Men seen in Singapore Town on Feb. 9 and 10th were heard to boast that they had come "down the line" because they were fed up with being plastered! When the S.S. Empire Star arrived at Batavia on Feb. 14th, several Australian deserters were taken ashore under guard. There have been allegations that men who fought valiantly in North Johore during the daylight hours walked back to a nearby township at night to buy beer! There were cases of looting and rape.[156]

Members of the British Navy also brought back reports of unruly and undisciplined Australian soldiers on Singapore Island.

Some of these reports were undoubtedly true and were made especially believable by the fact that the Australian 8th Division's commanding officer, General Gordon Bennett, handed over command of his division to his subordinates and commandeered a small boat just as the island surrendered. He eventually made it back to Australia, where he asserted it had been his duty to escape so that he could pass on the lessons he had learned. Few of his colleagues or subordinates agreed.[157] Thousands of Australian soldiers who remained in Singapore, along with their British and Indian colleagues, became POWs. They suffered in horrendous conditions for the next four years. Bennet's actions and his attitude prior to his departure clearly affected the soldiers around him, especially those subordinates who were left with the responsibility of looking after the 8th Division's soldiers in Japanese POW camps.[158]

Other reports acknowledged Australian indiscipline on Singapore Island but emphasized that the opposite had been true of Australian troops on the peninsula and that Indian and British troops were also demoralized and undisciplined once they reached Singapore. Ian Stewart of the Argyll and Sutherland Highlanders wrote:

> The A.I.F to my knowledge fought very well indeed. We were actually under the command of the Australian Division in the fighting on the island. It is true that a number straggled down to the town, and did make an early getaway, but this applies equally to British troops. There has been some talk that there was falling out between our particular regiment and the Australians. This is quite untrue. Of their behavior up country I have no knowledge. Admittedly a bad Australian is the worst thing ever, but a good one, to my mind, is the best.[159]

Additionally, notes from General Headquarters indicate that the staff there received only good reports of the AIF's motivation in Malaya and blamed the indiscipline on Singapore Island in part on the attitude of their commanding officers, made plain through actions such as General Bennett's escape.[160]

Except for a few small units, the 8th Australian Division did not join combat until January 14, 1942, at Gemas. However, that six weeks was not spent relaxing. Despite numerous battles to the north that demonstrated the importance of mobility and the limited benefits of fixed positions given

Japanese flanking tactics, the Australians continued to prepare fixed positions with which to defend the Johore region.[161]

Some unit commanders did realize the importance of mobile warfare and offensive tactics. The commanding officer of the 2/18th Battalion said in his Christmas message to his troops, "Although we have for the past few months been preparing more or less static positions all ranks must realize that static defence will not beat the enemy. The force that sits down will lose the battle, and we must at all times be prepared for aggression and a war of movement. The enemy must be hunted out and attacked on every occasion."[162] Those commanders who realized the value of mobile and jungle warfare attempted to instill those skills in their soldiers during the weeks before the battle reached them. While that training undoubtedly aided the soldiers once they met the Japanese, it also meant that the Australians were not completely rested. Digging, building, and training exercises filled their time, and the training contradicted the efforts put into preparing fixed positions. Nevertheless, they were in much better shape physically and mentally than the British or Indians by the time the remnants of those forces passed through the Australian lines and headed for Singapore.

Australian soldiers were eager for battle and demonstrated initiative, so much so that fifty of them volunteered to go behind enemy lines and target Japanese supplies and communications. Major Angus Rose, a member of the Argyll and Sutherland Highlanders who had been assigned desk duty in Singapore, came up with a plan to insert a battalion into Japanese-held territory to interfere with supplies and harass the Japanese from behind. Command could not spare an entire battalion but allowed Rose to seek out volunteers for a smaller mission, to be inserted by sea. Fifty Australian soldiers volunteered for the job.[163] They landed up the coast on December 27 and the next day successfully ambushed a supply convoy, killing a brigade commander in the process. For reasons unknown to the excited soldiers, command then recalled them and they returned to their units in Johore.[164]

The rest of the 8th Division had to wait until January for action and in the meantime witnessed the withdrawal of most of the British and Indian forces through their lines to Singapore.[165] On January 27 the Australian Infantry Brigade moved north from Johore to Gemas, and the 2/30th battalion went even further north to serve as a "bumper." They selected a bridge across the Gemencheh River to ambush the Japanese. According to the officers' reports after the battle,

> All through the night of the 13th the ambush troops lay in the mud and the pouring rain. No one was allowed to sleep and by mid-day of

the 14th the men were cursing the Japs for their tardiness. Stiff and cramped from lying in the mud and their wet clothes, their stomachs cold and unsatisfied from a thirty-six hour diet of cold bully beef without even the warming comfort of a cup of tea, their already eager enthusiasm to get to grips with the Japs was now sharpened by bad temper.[166]

Their patience was rewarded when Japanese cyclists began to cross the bridge. Waiting until the unit was split on either side of the river, the Australians blew up the bridge and ambushed the cyclists on their side. The ambush was successful and the 2/30th returned to their brigade with limited casualties.[167]

The Japanese quickly repaired the bridge the Australians had blown up (a common occurrence throughout the fighting on the peninsula) and crossed the river on January 15. Demonstrating initiative, the 27th Brigade engaged in patrolling, actively seeking out engagement with the Japanese, and attempting to prevent them from easily digging into new positions. Additionally, the Japanese tanks failed to inspire the same awe and fear in the Australian soldiers that they had in the British and Indian soldiers. Despite offensive patrolling actions and standing firm in the face of tanks, the 27th Brigade was severely outnumbered.[168] On January 16 it was withdrawn back to Johore after division headquarters began to fear it would be encircled and lost. According to the officers present, "good battle discipline had permitted the withdrawal to be carried out successfully."[169]

When the 45th Indian Brigade needed reinforcements at Muar River, the 2/19th and 2/29th Australian Battalions were sent to their aid. The Australians provided strong resistance, attacking the tank formations that had broken the Indian battalions and conducting a fighting withdrawal to try to help the 45th break free of what was quickly becoming a Japanese encirclement.[170] The 45th had been thoroughly cut off by Japanese forces, were out of communication, and could not coordinate their movements with the Australian units attempting to reinforce them. Both Australian units engaged in a "strong fight" as they attempted to reach the 45th, but the Japanese completed their encirclement. The 45th was forced to disperse into the jungle, and the Australian units withdrew south to Bakri.[171]

Meanwhile, further southeast, the 22nd Brigade—of which the 2/29th was a part—had also engaged the Japanese. After three days of unsuccessful fighting, the Australians once again demonstrated initiative and flexibility. The 22nd Brigade adopted guerrilla tactics, entering the jungle and attacking the Japanese from all angles. Not used to being on the receiving end of such tactics, the Japanese forces were confused, and disorganization delayed their progress south.[172]

Back at Bakri, the 2/29th Australian Battalion was attacked by Japanese tanks and infantry. Noticing that the tanks rushed ahead of the infantry—something the Japanese forces had been doing throughout the campaign—the Australians took the opportunity to try a new tactic. Waiting for the tanks to pass, the battalion surrounded the tanks and fired from all angles, taking out the crews as they abandoned their burning vehicles.[173] They were able to destroy eight tanks before withdrawing farther down the road to await the slower Japanese infantry.[174] While Japanese tanks had scared even the well-trained Argylls into retreat, the Australians engaged their better-equipped enemy long enough to exploit a tactical weakness. This particular battle, as well as the next four days' worth of delaying actions, kept the Japanese forces occupied long enough to begin the removal of all imperial forces to Singapore Island. By January 22 the retreat from the Malayan peninsula began in earnest. The Japanese general later described his unit's fight with the Australians south of Muar River as "the most 'savage encounter' of the campaign."[175]

Australian units demonstrated high levels of will to fight in Malaya. They displayed strong morale, with a good attitude and commitment to the fight. They practiced good discipline, observing orders and withdrawing when instructed to—even when they expressed a preference to stay and fight. Unlike either the British or Indian units, Australian units also demonstrated initiative. There were volunteers to go behind enemy lines and adopt patrolling and guerrilla tactics in an effort to adjust to the Japanese infiltration and encirclement tactics.

CONCLUSION

Given the material, strategic, and political limitations facing the garrison in Malaya, victory was never a likely outcome of the battle. However, this chapter has demonstrated that when faced with the same basic material circumstances and the same enemy, the three national militaries fighting in Malaya had different levels of will to fight. All three groups' will is best explained by national identity theory. British soldiers did not understand why they were fighting or view their battles as important to defending their nation. Indian national identities were not compatible with the goals of defeating Japan. Though a minority, some Indian nationalists even saw the Japanese as a potential ally in the struggle for independence. Indian soldiers did not identify with the British Empire and were not committed to its defense even when their homeland was in danger. Australia's national identity was closely connected to the goals of the war against Japan. Their white supremacist and imperial identity created a particularly high threat perception for Australian soldiers and motivated them to fight with strong will against the Japanese.

NOTES

1. TNA CAB 106/162.
2. Warren, *Singapore, 1942*, 6.
3. At this time, Britain held colonies in India, Burma, Malaya, and Hong Kong.
4. Kirby, *Singapore*, 25.
5. Kirby, *Singapore*, 26.
6. Smith, *Singapore Burning*, 57.
7. Kirby, *Singapore*, 109.
8. Kirby, *Singapore*, 110.
9. Kirby, *Singapore*, 111.
10. Kirby, *Singapore*, 51.
11. Bayly and Harper, *Forgotten Armies*, 107.
12. Kirby, *Singapore*, 113.
13. Kirby, *Singapore*, 110.
14. Moreman, *The Jungle*, 24.
15. Moreman, 24.
16. AWM 54 553/1/6.
17. Kirby, *Singapore*, 92; Roy, *The Battle for Malaya*, 42.
18. Kirby, *Singapore*, 96.
19. Kirby, *Singapore*, 94.
20. Kirby, *Singapore*, 96; Roy, 47.
21. Smith, 271.
22. As noted below, 12th Indian Brigade was made up of both British and Indian units.
23. Kirby, *Singapore*, 96.
24. Marshall, *To Have and Have Not*, x.
25. Darwin, *The Empire Project*, 463.
26. Raghavan, 68.
27. Churchill, *His Complete Speeches*, 6504.
28. Churchill, 6539.
29. Greenwood, *Why We Fight*, 22.
30. Greenwood, 108.
31. Gallup, *Gallup International Public Opinion Polls*, 25.
32. Gallup, 54.
33. Gallup, 55.
34. Grant, *Australian Soldiers in Asia Pacific*, 60.
35. Horne, *Race War*, 190.
36. Horne, 186.
37. Horne, 192.
38. As quoted in Smith, 269.
39. TNA WO 106/2550A.
40. Fennell, *Fighting the People's War*, 197.

41. Smith, 240.
42. TNA CAB 106/53.
43. Smith, 242.
44. Kirby, *Singapore*, 144.
45. Smith, 246.
46. Warren, 98.
47. Smith, 270.
48. TNA WO 106/2550A.
49. TNA WO 106/2550A.
50. Smith, 273.
51. TNA CAB 106/54.
52. Kirby, *The War Against Japan*, 247.
53. Smith, 279; Warren, 122.
54. TNA CAB 106/55.
55. Warren, 124.
56. TNA CAB 106/55.
57. TNA CAB 106/55.
58. Smith, 283.
59. Smith, 312 and 314.
60. TNA CAB 106/56.
61. The other units in the 12th Brigade were Indian and are discussed below.
62. Kirby, *The War Against Japan*, 276.
63. Smith, 329; Kirby, *The War Against Japan*, 277.
64. Kirby, *The War Against Japan*, 277.
65. Kirby, *The War Against Japan*, 278.
66. Farrell, *Defence and Fall of Singapore*, 215.
67. TNA CAB 106/55.
68. Smith, 346.
69. Warren, 137.
70. TNA WO 106/2550A.
71. Khan, *India at War*, 94.
72. Raghavan, 57.
73. Bayly and Harper, 15; Khan, 112.
74. Bayly and Harper, 5.
75. Raghavan, 11.
76. Raghavan, 278; Bayly and Harper, 19.
77. Bhattacharya, *Propaganda and Information*, 155.
78. Bhattacharya, 150.
79. Bhattacharya, 165.
80. Khan, 8.
81. Bayly and Harper, 76.
82. Bayly and Harper, 65.
83. Grant, 46.
84. Bayly and Harper, 66.
85. Kirby, *The War Against Japan*, 189.
86. Bhargava, *Campaigns in South-East Asia*, 135; Raghavan, 191; Roy, 86.
87. Smith, 165.

88. Smith, 165.
89. Smith, 150.
90. Kirby, *The War Against Japan*, 203–4.
91. Warren, 82; Bhargava, 148.
92. Bhargava, 148.
93. Warren, 83.
94. Bhargava, 149.
95. Bhargava, 149; Warren, 84.
96. Roy, 98.
97. Roy, 99; Bhargava, 151.
98. Bhargava, 151.
99. Smith, 240.
100. TNA CAB 106/162.
101. Warren, 84.
102. Smith, 239.
103. TNA CAB 106/54.
104. TNA CAB 106/54; Bayly and Harper, 66.
105. Kirby, *Singapore*, 144.
106. Kirby, *The War Against Japan*, 204; Bhargava, 143.
107. Kirby, *The War Against Japan*, 206.
108. Kirby, *The War Against Japan*, 206; Bhargava, 162.
109. TNA CAB 106/54.
110. Warren, 94; Roy, 99.
111. Bhargava, 163.
112. TNA CAB 106/54.
113. Smith, 245.
114. As quoted in Warren, 100; also described in TNA CAB 106/54.
115. TNA CAB 106/54.
116. Warren, 104.
117. TNA CAB 106/54.
118. Roy, 116.
119. Warren, 122.
120. Smith, 279.
121. TNA CAB 106/55.
122. Warren, 124.
123. Warren, 124.
124. TNA WO 106/2550A.
125. TNA CAB 106/162.
126. Smith, 314.
127. TNA CAB 106/54. This unit, unlike the 1st Hyderabad State Forces described earlier, was a part of the regular Indian Army and not on loan from the Hyderabad Maharajah.
128. Bhargava, 218.
129. Warren, 133.
130. Warren, 133; Smith, 328.
131. Smith, 329.
132. Roy, 136.
133. Smith, 332; Bhargava, 217.

134. Smith, 333.
135. Farrell, 218.
136. TNA CAB 106/55.
137. Warren, 160.
138. TNA CAB 106/162; Bhargava, 257.
139. Smith, 362.
140. Smith, 362; Roy, 150.
141. Warren, 176; Farrell, 280.
142. AWM SALT Volume 1.
143. AWM SALT Volume 2.
144. AWM SALT Volume 1.
145. AWM SALT Volume 1.
146. Welsh, *Australia: A New History*, 428.
147. Cantril, *Public Opinion*, 1072.
148. As quoted in Johnston, *Fighting the Enemy*, 88.
149. Grant, 26.
150. Grant, 32.
151. Grant, 41
152. Grant, 41.
153. Grant, 70.
154. Johnston, *Fighting the Enemy*, 78.
155. Jackson, *The British Empire and the Second World War*, 471.
156. TNA WO 106/2550A.
157. AWM 54 553/1/6.
158. AWM 54 553/1/6.
159. TNA WO 106/2550A.
160. TNA WO 106/2550A.
161. TNA CAB 106/162.
162. AWM 52 8/3/18/12.
163. Smith, 293.
164. TNA WO 106/54.
165. Smith, 355.
166. AWM 54 553/5/1.
167. Warren, 159; Roy, 153.
168. TNA CAB 106/162.
169. TNA CAB 106/162.
170. Smith, 366 and 375.
171. TNA CAB 106/56.
172. TNA CAB 106/56.
173. Warren, 164.
174. Warren, 166.
175. As quoted in Warren, 177.

CHAPTER 5

Europe

Although they all served in various engagements throughout Europe, British, Indian, and Australian forces did not fight together in any single campaign on that continent. To explore the goals of the war against Germany on the European continent and its relationship with the national identities of the three nationalities of interest, I instead selected one campaign for each nationality: the Australians in Greece (April 1941), the Indians at Monte Cassino, Italy (January to March 1944), and the British at Anzio, Italy (January 1944). This method increases the chances that confounding factors such as equipment or geography/topography may influence soldiers' will to fight, so I draw more cautious conclusions from the comparisons of each group to the other across the battles. However, these battles are useful for comparison within each group: for example, do the Australians have varying levels of will when fighting in Europe compared to North Africa or Malaya? Though there are more differences between the battles in these cases than in North Africa or Malaya, the results do provide additional evidence for the importance of national identity as well as some interesting implications for small-unit cohesion.

GREECE

Background

In February 1941, after having pushed the Italians almost entirely out of North Africa, the British sent imperial forces to Greece to help defend against a potential German invasion. The Greeks were holding off an Italian invasion, and intelligence sources indicated that the Germans intended to join their ally and attack Greece as well. The British government convinced the Australian and New Zealand governments to send their soldiers to Greece, and at the end of February and beginning of March 1941 British, New Zealand, and Australian units that had fought a successful campaign against the Italians in North Africa faced the Germans in Europe.

On April 12, 1939, Italy invaded Albania, prompting British Prime Minister Chamberlain to promise support to Greece should they too experience Italian or German aggression. The next day, in a speech in Parliament, Churchill voiced his own support for this promise: "I am also in agreement with the practical steps with which the Prime Minister has announced on behalf of His Majesty's Government to give a guarantee to Greece and to make even more effective arrangements with Turkey."[1] He continued to see that commitment as central to the British war effort, if only for the sake of appearing to honor alliance commitments.[2] As such, the decision to send imperial troops to Greece was a political one with faint hope for military success. Indeed, according to the official history, "The British campaign in the mainland of Greece was from start to finish a withdrawal."[3]

Initially, Australian leadership was ambivalent about the campaign. While in London, Australian Prime Minister Menzies sent a telegram to Acting Prime Minister Fadden in Canberra describing the plan as risky but saying that Churchill considered "the loss would be primarily one of material and that the bulk of the men could be got back to Egypt."[4] Australian officials would later claim that the British kept them out of the loop and oversold them on the chances of success. According to historian Maria Hill, "the Australian government and the press reacted very badly when news reached Australia that Australian troops had been placed in a defenseless position on Greece and Crete. Editorials scathingly critical of Britain's mishandling of the campaign were published with accusations of incompetence and poor leadership directed at both the British military leaders and Australian government."[5] Before the venture began, Menzies expressed frustration to the British Cabinet that they had been committed to the action without consultation, but said that "Australia was not likely to refuse to take a great risk in a good cause."[6] When he pressed Churchill on the possibility of success, Churchill replied that the "real foundation for the expedition was the estimate made on the spot of the 'overwhelming moral and political repercussions of abandoning Greece.'"[7] On February 26, 1941, Menzies telegrammed Fadden laying out the British position, concluding: "With some anxiety, my own recommendation to my colleagues is that we should concur."[8]

Military leaders also disagreed about the wisdom of the campaign. General Wavell, then commander of the Mediterranean forces, was in agreement with British political leaders about the decision, even helping to negotiate the plan with Greek officials.[9] However, Australian General Blamey and New Zealand General Freyberg expressed concern. On March 9, 1941, Blamey cabled the Australian cabinet asking permission to submit his own views in addition to Wavell's; this alarmed the cabinet both because the British had told them that Blamey had been consulted and because Blamey said that he believed the campaign to be most hazardous.[10] In his after-action report following the evacuation from Greece, Blamey wrote: "The outstanding lesson of the Greek Campaign is that no reasons whatever should outweigh military considerations when it is proposed to embark on a campaign, otherwise failure and defeat are courted. . . . As far as my limited knowledge goes the main reason for the dispatch of our force appears to have been a political one, viz., to support the Greeks and vindicate our agreed obligations."[11] Despite the concerns he expressed to his government before the campaign (though after the troop movements had begun), they approved the action.[12]

Blamey had good reason to be concerned. The units sent to Greece in April 1941 had just completed an intense campaign driving the Italians 500 miles from Mersa Matruh in Egypt to Benghazi in Libya between December 1940 and February 1941. The units involved in the fight were enjoying their victory but also worn out from it. Meanwhile some of the newer units had not yet been fully trained. The armored units that had participated required a complete overhaul, and much equipment had been lost.[13] Whatever functional equipment remained was fitted out for use in the desert and was not appropriate for use in the Greek mountains during a cold spring. There was not enough mobile transport, nor adequate roads. Soldiers were equipped with desert gear, not prepared for the mountainous conditions and snowy weather.[14] And there was never enough air support.[15] These poorly equipped imperial units were sent to reinforce Greek units who had been fighting the Italians for several months and were nearing their breaking point. It was under these conditions that Australian soldiers engaged the Germans in Greece.

Perceptions of the War

The war against Germany in Greece was pitched to the Australian people by both Australian and British leaders as imperative to helping the Greeks defend their homeland against German tyranny. The British and Australian governments portrayed the expedition to Greece as fulfilling a British promise to aid Greece against fascist aggression.[16] On April 9, as German troops were pushing imperial forces back across the desert in North Africa, Churchill gave a speech before Parliament outlining his reasoning for sending troops to Greece. "We were bound in honor to give them all the aid in our power. If they were resolved to face the might and fury of the Huns, we had no doubts but that we should share their ordeal and that the soldiers of the British Empire must stand in line with them."[17] He also stated that a "sound military plan giving good prospects of success" had been developed by the generals in the region, stretching the truth to the breaking point for propaganda purposes, something that would come back to haunt the government after the withdrawal.[18]

Calling on Australian national myths that emphasized spreading democracy, the Australians' mission in Greece was to defend the birthplace of democracy against a tyrannical foe. Newspapers promoted the idea, "that the British were reluctant to enter Greece but compelled to do so because of reasons of 'conscience' to support heroic little Greece's stand against the might of the Axis powers."[19] Hill notes that this was not an accurate assessment of the reasons for sending forces to Greece, but it was widely believed by Australians, civilian and military, at the time.[20] Additionally, Australian leaders called on the sense of connection to the British Empire and made the argument that holding Greece was necessary to holding the Mediterranean, and holding the Mediterranean was vital to British military power more generally and thus to Australia's own survival.[21]

At this point in the war Australian leaders (and AIF members) were growing concerned that the Australian public was not sufficiently concerned about the war or dedicated to the war effort. One military censorship report on the state of morale in mail from the military as a whole notes: "Frequently references are made in letters to strikes by Australian workers engaged in production of munitions and other essential wartime activities. The opinion usually expressed is that they are caused by the activities of Communists, with the support of Russia, and the government is criticized for not taking more drastic steps to stop the activities of such subversive elements. Other comment often noticed is that *the people are not putting all their weight into the war effort.*"[22] Concerned about the "apathy" of civilians toward the war, anthropologist A.P. Elkin ran a study in early-mid 1941 modeled in part on the British Mass-Observation project, surveying 400 Australians on their attitude toward the government and the war. He found that only 7 percent were truly "apathetic" toward the war itself.[23] Rather, lack of enthusiasm for the war could be traced to distrust of political leaders and lack of knowledge about the moral and political objectives of the war.[24] No survey asked about the Greece campaign specifically, but there was a significant public outcry after the defeat. Civilian censorship reports from May 1941 note:

> Greece and Crete: These two adventures continue to invoke an increasing volume of adverse comment and criticism expressed in personal letters of the public.... Much of the comment is to the effect that the politicians are responsible for the continued defeats, by interfering with the plans of the army leaders, writers stating that our troops should never have been sent without adequate equipment to enable them to meet the enemy on a basis approaching nearer to equality.[25]

This assessment lines up with Elkin's finding that negative attitudes toward the war were more closely related to distrust of political leaders than a lack of support for the war's goals.

Australian military leaders also sought to draw on myths and symbols of Australian national identity to motivate their soldiers. On April 12, 1941, the 1st Australian Corps (which included all of the Australian units present in North Africa and Greece) and the 2nd New Zealand Expeditionary Force (New Zealand's units in the region) were combined and renamed the ANZAC Corps. This move was designed by Blamey and New Zealand General Freyberg specifically to draw on the myths of the first ANZAC Corps in WWI, a force that played a central role in shaping Australia's national identity. According to officers' reports after the withdrawal, "the change was appreciated by all in the Corps."[26]

In addition to identifying themselves with the mythic heroic soldiers of WWI, Australian soldiers sent to Greece were upbeat due to their recent success against the Italians in North Africa.[27] Moreover, they saw themselves as helping to defend a democracy against the real enemy—the Germans.[28] According to one Australian captain's personal diary, Australian soldiers were "eager to come to grips with the Hun at long last."[29] Beyond finally being able to fight the true enemy (the mighty Germans as opposed to the laughable Italians), they were defending the birthplace of democracy against that tyrannical foe. "This race [the Greeks] is so like the British, every Hellenic soldier without his freedom would rather die—at last I have fully realized what we are fighting for."[30] Captain Oliphant claimed that he was not alone in feeling that they had a real chance to deliver a blow to fascism in this campaign.

The sense that theirs was a just cause was reinforced as Australians witnessed the Germans target civilians throughout the country.[31] Indeed, in addition to fighting to defend democracy, Australian soldiers also felt a racial affinity with the civilians they were defending. Throughout the North Africa campaign, racist attitudes toward Arabs were prevalent. In Greece, however, there was a sense that the country's white-European population appreciated the Australians' efforts on their behalf, unlike the Arabs among whom they had just served in Egypt and Libya. According to one journalist traveling with the soldiers, "It was grand to be in a place where you're welcome and where troops coming to fight against an invader aren't greeted with surly looks and regarded as fair game for a fat profit."[32] Oliphant also noted the fanfare with which they were greeted: "Our entry into Greece was like a Royal welcome—cheering crowds an excitable race the Greeks—who threw roses into our cars as we drove off to our first camp site."[33] Many Australian soldiers contrasted the clean, appreciative, and generous Greeks against what they viewed as dirty, thieving Arabs and were much happier to be defending the Greeks.[34] Australian discourse from North Africa speaks more of fighting in the area and around the population, while discussion of Greece references fighting for the Greek people. The importance of whiteness in Australian identity influenced soldiers' attitudes toward the people they were fighting among and the perception of the goals for which they were fighting.

The Campaign

The first major battle between Australian and German forces in Greece occurred April 11–12, 1941, at the Veve Pass—a major route south from Yugoslavia. The plan was to conduct a delaying operation in the mountains in the north of Greece to allow time to establish a firm defensive position at the Aliakmon River.[35] Mackay Force (named for General Mackay, the unit's Australian commanding officer) was sent forward to conduct that delaying operation at the Veve Pass. Mackay Force consisted of the 19th Australian Brigade in cooperation with the 21st Greek Brigade and the 1st Rangers, a British armored unit.[36] All the British and Australian units had traveled with little rest or re-equipment from North Africa to mountainous northern Greece. The 2/4th battalion (19th Brigade) and the Rangers arrived only on April 9, while the 2/8th battalion (19th Brigade) arrived on April 10 after a two-day march up the mountains that involved sleeping without shelter in the snow.[37]

The line was spread thin in an attempt to cover the road through the pass as well as the high ground around it. The Greek forces held the far right, the 2/8th next to them on the left, the Rangers with their armor sat astride the road, and the 2/4th was on the far left along the hills.[38] On the afternoon of April 11 German infantry approached the 2/4th's area but were stopped with artillery fire. The 2/4th lacked an artillery observer to call in fire, and so, demonstrating initiative, the unit's commanding officer directed fire via telephone through battalion headquarters.[39]

While the 2/4th was able to stop the German infantry attack, the 2/8th were also under attack on the right, with the Germans using infiltration tactics to move forward between the posts of the 2/8th battalion.[40] The 2/8th held off the attackers and even took several prisoners, on which they discovered that they had spent the night fighting members of the infamous SS unit the "Adolf Hitler."[41] The 2/8th's commanding officer was concerned about the possibility of infiltration, so the night of April 11 he ordered all troops to remain in their rifle pits and awake through the night: "You may be tired. You may be uncomfortable. But you are doing a job important to the rest of our forces. Therefore you will continue to do that job unless otherwise ordered."[42] This meant that the already tired battalion was even more exhausted by morning, when the Germans launched a heavy attack on the junction between the Rangers and the 2/8th battalion.

At this point it is important to note there is significant debate over the performance of the 2/8th battalion during the fighting on April 12. Once the Germans launched their full-scale attack, communication between the Rangers and the 2/8th was cut off. One platoon was overrun and as a result the Rangers believed that the 2/8th was withdrawing.[43] If that had been true, the Rangers' right flank would be in grave danger. One officer present with the 1st Armoured Brigade said that Greek and Australian troops were "swarming down the Amyntaon fork [as] ... Bosch had broken through the Rangers and the 2/8th Australians and the Dodacanese Div [Greek forces] had just gone."[44] As a result the Rangers pulled back to try to protect their flank.

Yet according to Gavin Long, the campaign's official historian: "After six hours of intermittent fighting in the pass and on the slopes to the east, the 2/8th still held the heights though their left had been mauled; the Rangers, however, were rallying astride the road about two miles to the rear."[45] The brigade commander ordered the Rangers to hold until dark and keep contact with the 2/8th, but the Rangers, thinking that the 2/8th had withdrawn, had already moved far to their rear. The 2/8th were now fighting the Germans on their own.[46] At this point the 2/8th was indeed overrun and was forced to retreat through the mountains, as the Germans now held the road.[47]

The level of organization of the 2/8th's withdrawal is contested. In an after-action report written soon after the withdrawal from Greece, Blamey wrote: "2/8 Bn mustered about 250 all ranks on the R. Aliakmon position the next morning [April 13] but of that number only about 60 were in possession of their arms. It is my intention to hold an enquiry into this position to ascertain how and why so many members of this Bn came to be separated from their weapons. This enquiry has been held up pending Brig. Vasey's return from Crete."[48] Hill suggests that the unit panicked and, in their haste to withdraw, abandoned their weapons.[49] It is true that many of the men were missing their weapons when they reassembled at the new battalion headquarters the next morning. However, others point out that the unit was very nearly cut off from the rest of their brigade and was without armored support when they began their withdrawal.[50] Moreover, the 2/8th battalion had not slept in three days and had spent the last seven days traveling and marching. Buckley argues "They had to get out quickly as best they could, and the only way of retreat was south-eastward and then down the road to Sotir. In this direction the survivors made their escape, most of them too exhausted to carry anything, so equipment and even arms had to be discarded."[51] In addition Long notes that Vasey, the officer who criticized the 2/8th in the records, later met with the commander and updated his assessment of the 2/8th's discipline under dire circumstances.[52]

While confusion reigned in the area of the 2/8th and the Rangers, the 2/4th held their sector on the left until 8 p.m. on April 12, as ordered.[53] They and the Royal Horse Artillery Australian anti-tank gun unit were able to keep the Germans from overrunning the rest of the brigade, allowing headquarters and other units to withdraw to Sotir.[54] The next day, April 13, the Germans caught up with the Australians and another withdrawal was necessary. The 2/4th, despite being below strength and having just fought a significant battle at Veve Pass, was required to fight the rearguard action to cover the withdrawal. According to their commanding officer, "there was not a murmur from any of the men of my tired battalion when told they had to fight this rearguard action."[55]

This first battle between Australian and German forces demonstrated how undermanned and under-equipped the Australians were but also saw Australian units fighting with initiative and discipline. Lack of communication between units caused confusion, but the 2/8th fought until they risked being surrounded and then withdrew through rugged terrain and reconvened with their brigade in the rear. The 2/4th held their sector as ordered, fighting against a better-armed German unit to allow their headquarters to withdraw, and then followed in an organized fashion. The 2/4th also demonstrated initiative in overcoming the lack of artillery spotters, developing a work-around to ensure that fire was called in as necessary.[56] Given the poor physical condition of the 2/8th and their lack of armored support once the Rangers withdrew precipitously, their withdrawal can be considered disciplined. All other units were able to withdraw via the road, while the 2/8th had to withdraw through mountainous terrain in the snow and still managed to rejoin their brigade.

From April 13 to 16, Allied forces withdrew south of the Aliakmon line to try to hold the mountain passes north of Larissa and the Plain of Thessaly to slow the German approach to Athens. The plan was to perform delaying actions there to allow time for Generals Wilson and Blamey to establish defensive positions at the new line of last defense—Thermopylae.[57] The goal was to hold the passes for three or four days and prevent the Germans from taking Larissa from the east and give the Allied troops to the north time to withdraw south to the new Thermopylae line.

The 16th Brigade had withdrawn south after the Veve Pass was lost and was now tasked with defending the Tempe Gorge, one of the main passes through the mountains northeast of Larissa. They had not fought any substantial battles with German forces but had been marching almost nonstop for the past week. They had destroyed or hidden much of their equipment when forced to retreat off road with only donkeys for transport, and so were under-equipped in addition to being exhausted when they took up their positions at Tempe Gorge.[58]

Because they had lost so much equipment, the 2/3rd and 2/2nd battalions (of the 16th Brigade) did not have enough signal wire to keep all of their units in touch, which meant runners were necessary to keep up communication between units and with leadership. The 2/2nd was deployed along the line, with the 2/3rd held in reserve.[59] The morning of April 18, elements of a German mountain division and an armored division attacked Tempe Gorge. One of the few mobile units in the area, a unit of carrier trucks, positioned itself on the left edge of the line along the river. Demonstrating initiative, the unit used the carrier trucks to intercept German troops crossing the river. Several men were wounded in the initial engagement, including their commanding officer, but the unit organized a rescue of the wounded men and established themselves on the left flank, preventing the Germans from encircling the 2/2nd Battalion.[60]

Further north, the 21st New Zealand Battalion was overrun.[61] At 11 a.m. small parties of the New Zealanders withdrew through the 2/2nd's line, telling the Australians that German tanks were ahead. At 3 p.m. the Germans attacked the 2/2nd forward platoons, beginning with a major air attack.[62] The Germans followed up the air attack with tank and infantry attacks, with infantry wading across the river between the 2/3rd and 2/2nd battalions.[63] The Australian carrier unit noted above slowed the advance across the river, but the Bren guns and machine guns were not able to stop the German tanks advancing down the road. Germans attacked the 2/2nd from the west, north, and northeast.[64] The 2/3rd withdrew south under orders, but the 2/2nd did not receive the order and was cut off from the rest of the 16th Brigade.[65]

A large number of German tanks and infantry now essentially surrounded the 2/2nd Battalion. At 6:45 a.m. the commanding officer decided that they could no longer delay the advance of the tanks and ordered withdrawal.[66] The unit withdrew under orders in an organized fashion into the mountains to the east but was cut off from the rest of the ANZAC corps and "ceased to function as such for the remainder of the campaign."[67] In his after-action report, Blamey was highly critical of the 2/2nd, writing that: "The action of 2/2 Bn withdrawing eastwards seems entirely due to the presence of 6–8 tanks on their left flank."[68] However, a note was later attached to the report correcting this statement: "The strength of the enemy forces opposing 2/2nd battalion at Penios Gorge on 18th April is not correctly given in 6th Australian Divisional Report. From later information received it has been confirmed that the enemy strength was considerably in excess of that shown in 6th Australian Divisional report, both in tanks and personnel, and was detailed in Anzac Corps report."[69] The 2/2nd were exhausted, under-equipped, and with only a carrier group for anti-tank support. They were able to delay a German tank unit for some hours before retreating under orders from their commanding officer. Additionally, the mobile carrier unit that provided support at the river demonstrated initiative and the ability to continue to function even after their commanding officer was incapacitated.

On April 16, as the British forces retreated south toward the Thermopylae line, the head of the Greek military, General Papagos, suggested to General Wilson, the head of all the overall imperial force, that to prevent his country's destruction, the British should withdraw and Greece surrender. A few weeks before, on March 30, German forces had attacked the imperial Desert Army in North Africa, and by this point British imperial forces had retreated as far as the Libyan-Egyptian border and the 9th Australian Division (with some British and Indian units) was under siege in Tobruk in Libya (see chapter 3). The Royal Navy was stretched thin keeping the garrison at Tobruk supplied, and no reinforcements were available for Greece. General Wavell flew to Athens for consultations on April 19, and by April 21 had agreed with General Papagos: the imperial forces should withdraw from Greece. The defense of the Thermopylae line was now to be a delaying action intended to allow as much of the ANZAC and British forces to be embarked from the beaches and ports as possible.

Some rearguard actions occurred April 20–21 as imperial forces withdrew to the south and the Thermopylae line. The fighting retreat demonstrated the Australian forces' ability to withdraw in an organized fashion under fire and to continue to inflict loss on the enemy even after it was clear that the German forces would be victorious. The 2/6th and 2/7th Battalions encountered German motorcycle outfits on the Lamia-Damakos Road. They drove back the motorcycles and also put the armored car that followed out of action.[70] Major Savige notes in his report on his unit's withdrawal from Larissa to the Thermopylae line that it was conducted under serious air attack. "The tps [sic] were very tired and the intense bombing was effecting [sic] them."[71] He notes that while some drivers abandoned their vehicles earlier than absolutely necessary or sought cover too far from them (slowing the withdrawal and congesting the road), that overall "in the circumstances the behavior of the Tps [sic] was good."[72]

At Thermopylae, Australian forces were to defend the Brallos Pass, which included a road and train tunnel. Brigadier George Vasey was commanding but had not yet been told of the decision to evacuate from Greece. On April 20 Vasey gave the order: "Here we bloody well are and here we bloody well stay."[73] One unit under these orders, the 2/2nd Field Regiment, was positioned on the hill overlooking the demolished bridge over the Sperkhios River. Their two artillery guns were positioned to prevent the Germans from crossing the river. On April 21 the Germans approached and were quickly halted by the Australian guns, retreating back toward Lamia.[74]

Throughout the night, the 2/2nd watched lights in the distance as the Germans moved south from Lamia toward the Sperkhios. The next morning there was an exchange of artillery fire as the Australians fired on the Germans to keep them back from the river, and the Germans attempted to dislodge the 2/2nd. By 1 p.m. only one of the Australian guns remained functional, and it was leaking oil. The Germans were able to come all the way forward and begin unloading infantry. Demonstrating initiative, the 2/2nd "lifted the tail of the gun on to the edge of the pit so as to depress it enough to fire down the face of the hill and, using a weak charge lest the recoil should cause the gun to somersault, fired more than fifty rounds into the enemy infantry."[75] This drew further fire from the German artillery, forcing the Australians to take cover. When they returned to their gun it was destroyed. They spiked the guns, collected identity discs from the dead, and withdrew south.[76]

On April 23 the order to evacuate came from Wilson but included an order to defend the road to Athens until April 26 so that the British Navy could evacuate troops.[77] On April 24 the 2/11th Battalion was deployed astride the main road to Athens. Machinegun positions established by the 2/2nd Regiment had been bombed by air the night before, so that regiment had moved the guns and then turned over the position to the 2/11th. The 2/11th allowed the Germans to bomb the old gun positions again the morning of the 24th, manning their new gun positions and preparing for the German advance. At 11:30 a.m. machine gunners attached to the 2/11th began firing on the advancing infantry, and both sides exchanged mortar fire and machinegun fire until that afternoon. By 8 p.m. the German attacks tapered off and the German advance was halted for the evening. According to Long, "This was the second time within a fortnight that German armored forces, hitherto seemingly invincible, had been halted with heavy loss by resolute infantry and artillery [the first time was at Tobruk on April 11]."[78] The 2/11th proceeded to withdraw to Athens under orders to evacuate.

Meanwhile to the west, on April 23 the 2/6th Battalion (along with New Zealand and British units) was defending the canal at Corinth, protecting the troops evacuating from the Peloponnese peninsula. Savige notes in his description of this action that, "The perfect timing of all units, in the face of these difficulties over a front of 6 miles is, I consider, a monument of splendid leadership on the part of Wrigley and Walker, excellent control by all regimental officers and splendid discipline on the part of tired troops."[79] Throughout the day on April 26 the area was attacked by German planes, attacks that began again first thing the next morning. After the initial round of bombing and machinegun attacks, a new type of plane appeared and German paratroopers began to jump.[80] The Australians immediately opened fire on the parachutes and were able to shoot some of them down, but many were out of range and landed safely. Several Australian units engaged in close combat with the paratroopers and were overrun, while others withdrew and attempted to harass the German forces as they did so.[81] Their officer reported the withdrawal was carried out in good order.[82]

Company headquarters was also attacked by German paratroopers. The officers and clerks stationed there fought with determination for several hours: "Several grenades were thrown into the trench which the headquarters occupied but were thrown out again by Private Coulam, the company clerk, until one exploded in his hand and wounded him seriously in the face."[83] The Australians finally capitulated when they ran out of ammunition.

The parachute attack succeeded in cutting the imperial forces on the peninsula into two: "First, the main force from Argos southwards; second, the rearguard forces—the 4th New Zealand Brigade and the 1st Armored Brigade detachment—astride the roads north-west of Athens, and the artillerymen awaiting embarkation at the Marathon Beaches."[84] The dispersal of imperial forces across southern Greece combined with the lack of transport ships to remove them all led New Zealand General Freyberg to label the situation on April 27 "chaotic."[85]

The final significant battle took place at Kalamata, on the south-west of the Peloponnese peninsula. By April 26 18,000–20,000 imperial troops were assembled there and were slowly being evacuated by sea. About 10,000 were taken out that evening, with the remaining 8,000 waiting to be removed the night of April 27. About 4 p.m. that afternoon a British unit patrolling north of Kalamata reported that they had seen the enemy some twenty-five miles out. Two hours later the Germans overran the British patrol and moved into town. Fighting broke out around the harbor though the imperial forces were vastly outgunned, having destroyed or buried much of their kit during the day to prepare for evacuation.[86]

Demonstrating initiative in seeking out engagement with the enemy, 70 of the 400 Australians under command of Captain Gray were able to arm themselves. He divided them in two and joined the New Zealanders attacking in the harbor and along the beachfront. They were able to recapture the harbor and even took one hundred German prisoners.[87] Unfortunately, the ships that had been headed back to remove the remaining troops at Kalamata saw the fighting and, having been told that the Germans had captured the embarkation areas, turned away. The imperial forces had to surrender to the Germans.[88] According to one soldier who escaped Kalamata, "The Brigadier gave every officer and man the opportunity of escaping if possible, though the chance of escape were remote."[89] A number of soldiers opted to take the option, eventually making their way to Turkey and rejoining the imperial forces in North Africa, while others surrendered to the Germans.[90] The total imperial losses to the Germans during the campaign were 1,160 killed, 3,755 wounded, and 345 missing, with nearly 14,000 taken prisoner.[91]

The Australian units that fought in Greece had high levels of will to fight. They spoke and wrote positively of their mission at the beginning of the campaign. After the campaign, many wrote home expressing frustration at not having been given adequate resources but did not argue the point of the campaign.[92] One soldier wrote home, "Well, I suppose there are quite a lot back home with their tails between their legs because we evacuated Greece. You can tell them this—the Hun as a soldier, man to man, even with his tanks, is no match for ANZAC or Tommy. It was lack of air support together with fifth columnists."[93] There were no reports of desertion, and officers reported good attitudes and performance from their soldiers despite difficult physical conditions. The 2/8th comes in for criticism for indiscipline, but that claim is contested, and there is evidence to suggest that given the conditions and the progress of the battle, they in fact maintained discipline. There are also numerous examples of initiative as soldiers adjusted to their limited resources, making creative use of guns and trucks not designed to be anti-tank weapons and seeking out engagement with the enemy.

ITALY—INDIANS AT MONTE CASSINO

Background

The Second and Third Battles for Monte Cassino took place from January to March 1944, as part of the Allied march on Rome. After pushing the Germans and Italians entirely out of North Africa in May 1943, the Allied forces turned their attention to Europe.

At first the American 7th Army and the British 8th Army met with success, capturing Sicily in August 1943 and landing on the mainland on September 3, 1943, the same day the Italian government surrendered. Unfortunately, rather than simply taking Italy out of the war as military leaders had hoped, German forces in Italy dug in to defend against the invasion. Even then British and American forces made progress north until the end of 1943 when they reached the Winter Line—a series of heavily fortified German defensive positions combined with the rivers and mountainous terrain of central Italy.[94]

Monte Cassino, a town surrounding a large monastery perched on a mountain, formed a key part of the Winter Line. German paratroopers had occupied the town, fortifying the buildings and key positions on the mountain leading up to the monastery. The position itself was fairly small, a battlefield of about five miles north to south and four miles east to west.[95] The American 5th Army, along with the French Expeditionary Corps, attempted to flank Monte Cassino by crossing the Liri and Garigliano rivers downstream. These crossings failed and American General Mark Clark decided to attack the monastery head-on.[96] To try to spread the Germans more thinly, British General Alexander ordered an amphibious landing at Anzio Beach on January 22, hoping to outflank the German position at Cassino and cut their lines of communication with Rome.[97] The initial landing succeeded, but Allied forces were unable to break out of the beachhead (see the final section of this chapter for a more detailed discussion of Anzio). The Germans were able to reinforce their defenses both at Anzio and Monte Cassino, holding off Allied forces.

The first Battle of Monte Cassino began in mid-January and petered out by February 7, when British 8th Army forces arrived to relieve the Americans. Allied forces had captured several important points on the mountain approach to the monastery but had been unable to take either the town or the monastery itself. According to one historian of the battle, "Clark, as we have seen, drove II Corps to the very limits of human endurance to achieve this, desperately hoping, in the best traditions of the Somme and Passchendaele, that 'just one more push would do the trick.'"[98] By the beginning of February it was clear that the approach was not working, and the decision was made to bring in the 4th Indian Army Division and 2nd New Zealand Division (both part of the British 8th Army) to relieve the American forces and attempt a new assault on the position. The 2nd New Zealand Division would attempt to take the town from the Liri Valley to the southeast, while the 4th Indian Division would approach the monastery from the mountain positions to the northwest.

Beginning the second battle of Monte Cassino was imperative in order to take pressure off the Allied forces who had landed at Anzio. Initially intended to divert German forces away from the battles at Cassino, a new battle at Cassino was necessary to divert from the battle at Anzio. Because of the failures of the initial assault, American and British leaders gave in to the demands of commanders on the ground and agreed to an aerial bombardment of the monastery. It is here that we pick up the story of Indian forces fighting the German Army in Europe.

Perceptions of the War

Indian forces at Monte Cassino were continuing the battle against Germany they had begun in North Africa. As noted in chapter 3, the war goals the British articulated were not compatible with the diversity of national identities in India. British leaders portrayed the war against Germany as one to defend democracy, protect small nations, and champion the oppressed. Given Britain's own treatment of India, such proclamations seemed cynical and were rejected by Indian nationalists of many stripes.

Over the course of 1942–43, Indian nationalist attitudes toward the war hardened due to three major events—the Japanese attack in the east, the Cripps Mission, and a growing famine across India. Since chapters 3 and 4 already addressed the impact of Japan's entry into the war, the following discussion focuses on the famine and Cripps Mission. Indeed, throughout the censorship reports for 1943 and 1944, the famine is seen as by far the most imminent threat to India, both by civilians at home and soldiers abroad.

In March 1942 the British Labour politician Sir Stafford Cripps arrived in India bearing a proposal for India's constitutional status, intended to settle the issue so that the British could more easily draw on Indian resources in the now two-fronted war against Germany and Japan. The proposal promised to include Indian ministers in the Indian Cabinet (but not in the defense position), create a committee to write a constitution for dominion status after the war, and give each Indian state the option to "opt out" of that constitution.[99]

Congress and Hindu nationalists rejected the proposal because it paved the way for "Pakistan," in that it allowed Muslim majority areas of India to secede from India as a whole. Additionally, they objected to arrangements for shared governance during the war, especially the British refusal to give Indian ministers authority over defense issues while still involving nationalist parties in the conduct of the war as a whole. Gandhi worried over abandoning nonviolence, while Nehru and others were concerned about asking Indians to fight against the Indian National Army then being assembled by Japan while having no say in the conduct of that war.[100] Scholar of British colonial politics Nicholas Owen quotes Nehru as saying, "Nothing could be more dangerous . . . than to be saddled with responsibility without complete power."[101] Once Congress rejected Cripps's proposal, the Muslim League did so as well. While Congress worried that it gave Muslims the chance to secede from India, the Muslim League worried that it did not offer Muslims enough protections.[102] According to historian Yasmin Khan, "To put it plainly, supporting the Allied war effort had simply become too politically unpopular and risky in India."[103] Indeed, the Indian government's situation report to the War Cabinet in London for the time notes, "Anxiety about the war would have been greater but for the interest taken in the mission of Sir Stafford Cripps."[104] The mission was of great interest to all Indians, not just the nationalist politicians.

Once Congress rejected the British proposal, they began planning for a new round of actions designed to highlight British dependence on a subservient India—the Quit India movement. On April 27, 1942, the Congress Working Committee took up a resolution demanding that the British "Quit India," although Nehru succeeded in removing the clause demanding removal of foreign troops, concerned that India would not be able to defend itself from German and Japanese fascists without the British and American troops then present in the country.[105]

Indians were generally concerned about fascism even while opposing continued British rule. Despite adopting the resolution, Congress hesitated to implement the Quit India movement, which would encourage work stoppages and noncooperation with the war effort, until the threat of Japanese invasion receded. By late summer the danger had subsided, and on August 8 Gandhi gave a speech launching Quit India. The next day the British arrested Gandhi and the rest of Congress's leadership. Jinnah, leader of the Muslim League, specifically called on Muslims not to take part in the demonstrations.[106]

This concern over defeating fascism also shows up in letters from Indian soldiers, although they are generally dismissive of the Japanese threat to India itself. One soldier wrote, "Who would not dedicate himself, heart, soul and body to our Lord the King, and fight to free posterity from the power of the tyrants. To serve in this war is the duty of every able-bodied man, and we shall play our part and return as brave lads of mother India.... God save the King!"[107] This particular excerpt highlights the fact that not all Indians rejected imperialism. Yet they also identified with "mother India" and saw themselves as fighting against tyranny. An Indian officer, on arriving in Italy in October 1943, wrote home: "We are the first Indian formation to land on European soil in this war and feel very proud of it too."[108] Though not accurate, as Indian soldiers had been present in the first battles against the Germans in France, this is yet another example of Indian soldiers identifying specifically as Indian yet also seeing themselves as participating in a worthy fight. Referencing an ancient Sanskrit epic, a soldier in the 10th Indian Infantry Brigade described his long time away from home: "You need not worry about me in the least. It was only the will of God that I am suffering. It is banishment for me which is just the same as Rama and Sita."[109] Indian national identity, especially for those who volunteered to join the British Indian Army, was complicated. Indians identified with the trappings of empire but also saw themselves as distinct from it. It is also important to note that statements in support of Indian nationalism could be punished if found, so soldiers were not free to express all forms of identity in their letters home.

When Quit India demonstrations, work stoppages, and acts of sabotage spread across the country, the British instituted mass arrests and violent repression.[110] Khan argues that the movement was not simply directed by Congress but varied widely across India: "In towns and cities it was championed by prosperous students who led strikes, *hartals* and demonstrations. In the countryside, it was far more nebulous but also far more violent from the start. Pre-existing peasant activism and grievances fused with the movement, and at times undefined general crime and disorder segued into more organized political protest."[111] Though not a mass uprising, these actions across India demonstrated significant resistance to continued British rule and to the war effort, with demonstrations aiming to disrupt military transportation and even, in at least one case, threatening soldiers themselves.[112]

As the unrest died down, the impending famine began to make itself known. Food and kerosene prices started to rise. The British knew when Burma fell to the Japanese that food would become a problem—India imported large amounts of rice from Burma, imports that ended with the British scorched-earth retreat and Japanese occupation. There is a large literature on the causes and responses to the famine that started in Bengal but affected all of India, though scholars point primarily to poor harvest and massive British mismanagement.[113] Beginning in early 1943, the prices of all basic necessities shot up—food, kerosene, even fabric. Suspicion of government mismanagement was widespread.[114]

By early 1944 the inquiry set up to investigate the causes of the famine "described the conditions as 'a complete crack-up of rural life—rural economy, rural society, and rural humanity—this is the reality of 1944. The village artisans; weavers, fishermen, boatmen, black-smiths—all have been ruined. Fishermen have no fishing nets, boatmen have no boats, weavers get no yarn, black-smiths get no iron or steel.'"[115] According to estimates at the time, 10 million in Bengal were homeless and another 3.5 million had died.[116] Hunger drove people from Bengal to other parts of India that also experienced food shortages and the accompanying diseases of starvation, such as cholera. This crisis intensified communal tensions, as Hindus and Muslims accused one another of causing the famine and of misappropriating and misallocating relief supplies.[117]

The effects of the turmoil in India did not go unnoticed by Indian soldiers abroad. Censorship reports indicate that despite British efforts to keep news of Quit India and the famine from the soldiers, they knew and were affected. These reports should be read with care. While the censors were charged with giving an honest assessment of soldiers' morale, they read their correspondence as British officials participating in a colonial project. In several instances discussed below, censors judged morale to be high while also quoting, as representative, soldiers saying they were anxious and worried for their families. One wrote, "I read in the papers about the Bengal famine. It is so shocking that I can't express in this letter and we are so sorry that we are so small and helpless."[118] In addition, disciplinary action was taken against Indian soldiers based on the content of letters. Claims of discrimination were investigated, but so too were political statements and possible affiliation with nationalists.[119] Censorship reports provide a useful window into what soldiers and their families discussed in letters they knew would be read by British military officials but should be read with that power structure in mind.

As did soldiers in all the imperial armies, Indian soldiers in the 4th Indian Division kept up with the progress of the war around the world. One report dated July 14–27, 1943, noted, "The Russian stand against the latest German offensive has earned much praise though it is felt Russian successes are exaggerated."[120] Following 8th Army success in driving the Germans and Italians from North Africa, King George VI visited allied troops in Libya. Indian soldiers wrote home of seeing the "King Emperor" and the regimental pride it engendered: "H.H. the King came down to visit us and in Tripoli we had a very grand and most impressive parade of our army. It was indeed a godsend opportunity to see him. He was awfully pleased with the extraordinary gallantry of our army."[121] The censors reported no negative comments toward the king in the weeks surrounding his visit. Rather, numerous soldiers wrote home of their pride in the 8th Army, the 4th Division, and their individual units. There was some frustration among non-Gurkha units with the amount of press the Gurkhas received, but otherwise little in the way of negative comments regarding the progress of the war and the units with which they served.

Once they landed in Italy, religious identity manifested in various ways. One British captain of an Indian battalion wrote: "All our classes except the fanatical Pathan have got religion just now, something to do with a longer analysis that good God's warriors were largely recovered from enemy hands, while the less good are still POW. So the drums and cymbals used to sound full blast at all the moments in the mandar and gurdwaras and the P.M.s have very frequent and largely attended prayers at short intervals, finishing with a fully choral performance in the evenings."[122] It is not clear that this officer fully understood the religious practices of his soldiers but was aware of the increasing salience of religious identity among them.

Moreover, some apparently had religious interpretations of the fight in Italy itself. One Indian soldier, writing in Urdu, said: "We are now part of Central Mediterranean forces and have advanced far beyond. Now we are in the country of the one who is an old enemy of the Muslims. Though we are facing great dangers but we have stout heart and shall go forth boldly till the last minute and have complete trust in God and remember that we shall gain our end in spite of everything the enemy does."[123] The importance of religion may have been due in part to the fact that the Italian civilians they were now living among clearly identified Indian soldiers as different. One Sikh subaltern wrote: "During the first week after our arrival civilians were very much frightened to see our troops (especially Sikhs). They had the idea that Indians kill children to eat, isn't it very funny."[124] Religious identity informed soldiers' sense of their relation to the war, to each other, and to the people they were fighting for and against.

Not only did Indian soldiers keep up with war news and thinking about their own part in that war, but they were also engaged with the increasingly dire political and economic situation at home. Immediately following the Cripps Mission in 1942, the Indian government received unit intelligence stating that "little or no interest in the proposals has been shown by the Indian armed forces."[125] However, the report goes on to note that Gurkha soldiers would reject the proposal as it would endanger their status in the Indian Army postwar. Of Indian officers, "A number of these must be nationalist and would alone have welcomed a settlement on the lines of the Cripps proposal."[126] While denying soldiers' interest, the intelligence reports still indicate that soldiers "would" have opinions about India's constitutional status.

Even the censors could not deny the intense interest that Indian soldiers had in the economic situation at home as the famine developed in the fall of 1943. Censors attempted to keep the worst of the reports from the soldiers in North Africa and Italy, but as the famine spread, they could not hope to be successful. "Many alarming reports concerning conditions in Bengal and S. India have had to be suppressed here to prevent the spread of alarm and despondency. The number of letters containing such reports is however on the increase and some letters must inevitably get through since letters emanating from places like Simla and Lahore are giving the news they get from Indian newspapers, which do not seem to be over reticent in the news they publish."[127]

It is telling that while claiming that morale was high, censors include the following extract from a soldier in 4th Division Head Quarters:

> Our comforts are very well looked after. In short, we are kept simply spick and span at present . . . but we are greatly worried over the increasing dearness and ultimate starvation of our relatives in India. Every one of us is perturbed. Our minds are full of anxieties. Many persons are returning back from India and they tell of the wretched conditions prevailing there. This is so disheartening. We cease to feel any rest and comfort though our conditions are so good when we remember how you civilians there, are undergoing the sufferings.[128]

This concern begins to show up in the letters Indian soldiers wrote home as early as June 1943 and continues throughout the period of the Battles of

Monte Cassino in early 1944. Not only is there anxiety on behalf of their families but also frustration with the government and the military. Families of soldiers were supposed to get easier access to food and necessities but were constantly writing to soldiers asking them to petition their officers to get that access. One officer wrote, "I have come to know that there is a scarcity of kerosene oil and sugar in India and it is controlled by the government. There are only small children behind at my home. I doubt if anybody helps them to get these articles. This arrangement is particularly made for the dependents of those military men who are on active service. I am anxious to know whether or not my family is supplied with these articles under control."[129] Soldiers who had been promised their families would be taken care of while they were away began to doubt the government's promises.

Indian soldiers went into battle at Monte Cassino proud of their units' success in North Africa but deeply concerned for the lives of their families at home. They doubted the government's promise to take care of their families even as they were asked to enter into brutal mountain battles to fight for freedom from fascism thousands of miles from home.

The Campaign

The 4th Indian Division moved to relieve American forces around Monte Cassino beginning in early February 1944. The initial plan for the Second Battle called for 4th Division units to be in place by February 12, ready for a new push toward the monastery by the night of February 14.[130] This timeline quickly fell apart. First, German units held the high ground, meaning no movement was possible during daylight. Second, once Indian units did move into position, they had to help the Americans fight off a counterattack from German forces. In fact, the Americans did not actually possess some of the key defensive positions that their commanders believed they did, meaning the attack would not even begin from the planned start points. Finally, some of the American units were so decimated by exposure and hunger that they had to be essentially carried off the mountain by Indian soldiers.[131] This delay meant that the air attack that had to take place on the 14th due to weather and resource concerns would precede the ground attack by nearly two days.[132]

The ground attack on Monte Cassino launched the night of February 17. The units involved were the 1/9th Gurkha Rifles, the 1/6th Rajputana Rifles, the 4/16th Punjab Regiment, the 1/2th Gurkha Rifles, the 4/6th Rajputana Rifles, and the 2/7th Gurkhas. Additionally, two British units were also involved—the 1st Royal Sussex and the 1/4th Essex, all part of the 4th Indian Division, which had successfully engaged in battles against German and Italian forces throughout Africa and Italy. Censors noted that these members of the 4th Indian Infantry Division had a strong sense of esprit de corps stemming from their successes.[133]

The air bombardment of February 14 blew the roofs off many of the monastery buildings, also collapsing some walls. Yet the majority of the buildings remained standing, and the damage did more to provide cover for German troops who now entered the monastery in force than to drive them out.[134] The two-day delay in launching the ground attack allowed German forces to settle into their new cover.

The night of the 17th, the 4/16th Punjabis and 1/9th Gurkha Rifles made some progress against German defenses but could not proceed much further without the capture of Pt. 593.[135] Previously believed to have been held by American forces, Pt. 593 was a crest on the mountain overlooking a major path toward the monastery itself. German possession meant that any forces moving up the mountain were exposed to sniper fire and bombardment from above. Several companies of 4/6th Rajputana Rifles had made some progress toward Pt. 593 but could not capture the crest itself. By dawn 4/6 Rajputana Rifles as well as the 4/16th Punjab settled in to hold their positions, knowing no progress could be made in daylight while the Germans possessed the crest.[136] The 1/9th Gurkhas, in the meantime, had attempted to advance despite the fact that Pt. 593 was still in German hands. They were able to advance as far as the southwest corner of the monastery but were driven back by German cross fire from the monastery itself and Pt. 593, driving them back and inflicting ninety-four casualties on the unit.[137]

Elsewhere on the line, the 1/2nd Gurkhas moved forward with their own attack on the monastery, aiming at Pt. 445 on the northern side. This required them to proceed through an area covered in shrub, which proved to be chest-high and covered in thorns.[138] As two companies of the 1/2nd Gurkhas were slowed, the German forces occupying Pts. 450 and 445 above them opened fire, reducing the two companies of Indian Army soldiers to a single platoon each within minutes.[139] Despite this destruction, some of the soldiers made it through the thicket and captured a stream in front of their objective.[140] The British official history notes that the 1/2nd Gurkhas lost 149 officers and soldiers in this attack.[141] Despite their success in destroying some of the German positions and capturing a stream on the way to Pt. 445, the remaining soldiers were too few and too injured to hold the position during daylight. Demonstrating discipline, they followed orders to withdraw and dig defensive positions.[142]

The 4th Indian Division put in a strong effort to take the monastery at Monte Cassino but succeeded only in moving the Allied lines slightly farther up the mountain. Numerous factors worked against them. First, the geography of the mountain prevented Allied forces from massing numbers against the Germans. Mountain paths meant that attacks had to proceed in small numbers and movement could occur only at night. Second, the air attack intended to clear supposed German snipers from the monastery only served to provide more cover and attract more German forces into the position. Finally, the supporting attack by New Zealand forces on Cassino town from the southeast had been meant to split German forces. However, the attack quickly bogged down in wet ground defended by Germans from entrenched, high positions. Indian units fought with discipline but made little progress under these conditions. Efforts to take the monastery were called off on February 18.

The next attempt to take Monte Cassino began on March 14. Between these two battles, 4th Indian Division continued to hold positions on the mountain, suffering continuous attacks by German forces as well as exposure to sleet and snow. Many units ran low on ammunition and food as they could only be supplied by porters traveling up the mountain trails at night.[143] A thorough reorganization of the front took place during this period, with fresh units from the 5th Indian Brigade (part of the 4th Indian Division) concentrated in a village called Cairo, located at the base of the mountain. The plan was for the 5th Brigade to attack the monastery from the northeast while the 2nd New Zealand Division, with its armored units, took Cassino town to the east. Having captured Cassino town, the armored units would then support the 4th Indian Division's infantry attack on the monastery. The whole battle would be immediately preceded by a thorough bombing attack on both town and monastery.[144] The plan was to begin this next offensive in late February, but the winter weather that struck on February 23 prevented aerial bombing, in turn postponing the ground attack. The delays and miserable conditions caused some decline in morale among the soldiers and drained their numbers as casualties from German harassing fire and sickness grew.[145]

Before this plan began, there were a few attempts to improve the 4th Indian Division's starting points. On February 23, the 1/9th Gurkha Rifles attempted to capture Pt. 445, which overlooked the 4th Division's planned approach to the monastery. Despite an artillery barrage to cover them, the fire from the German machinegun post was too strong to progress and the attack was called off.[146] The decision was made to proceed with the full offensive without taking Pt. 445.

When the bombing finally commenced on March 15, it did so with little notice to 4th Indian Division units stationed on the mountain. The 4/16th Punjabis had three bombs fall within fifty yards of their headquarters.[147] Once again, the bombing did not have the desired effect on German troops stationed in the town and monastery. The long delay in the attack had allowed time for German paratroopers to enter the position as reinforcements, and all German troops had hardened their positions with concrete and digging.[148] The damage done by the bombing made roads in town impassible to the New Zealand armored forces but created even more defensible positions for German forces.

On the night of the 15th, the attack up the mountain commenced, with "constant clashes with small enemy parties and posts" making slow going.[149] The Essex Regiment took Castle Hill at 3 a.m. on March 16, while the 1/6th Rajputana Rifles moved forward of Castle Hill to attempt to take Pt. 236. Cross fire was intense. One company withdrew to Castle Hill, but the other disappeared.[150] According to the Official Indian Army history, the unit's headquarters was struck by artillery shelling, killing nearly all of the officers.[151]

At the same time, the 1/9th Gurkha Rifles marched five hours from their base in Cairo Village to the outskirts of Monte Cassino. The unit's officer, Colonel Nangle, was unable to get information about other imperial forces in the area, so he decided to make an attempt on his goal of "Hangman's Hill" themselves. Hangman's Hill (Pt. 435) was positioned southwest of the monastery, between the monastery and Cassino town itself. It was considered a key position, control of which would protect forces approaching both the monastery and the town. "D" Company was pinned down by enemy fire from Cassino town, losing fifteen men within a minute.[152] Progress checked, Nangle believed his forces would not reach their objective before dawn. Unwilling to risk a daytime assault on the position, he withdrew back to the outskirts of Cassino town. He had, however, lost "C" Company in the process. No one would know this for several hours, but having lost communication with the rest of the Battalion, "C" Company had proceeded to Hangman's Hill and taken control of the position.[153]

On the morning of March 16, therefore, the 1/4th Essex and 1/6th Rajputana Rifles held Castle Hill. "C" Company of 1/9th Gurkhas held Hangman's Hill. "D" Company of 1/9th Gurkhas was on the outskirts of Cassino town.[154] Not knowing the Gurkhas held Hangman's Hill, Brigadier General Bateman ordered the Rajputana Rifles to take two important positions on the way to Hangman's Hill (pts. 236 and 202). He would then send the remaining 1/9th Gurkhas to capture Hangman's Hill. By 9:30 A.M. this plan had foundered when the Rajputana Rifles came under intense fire from Pt. 236 and withdrew under orders to Castle Hill. Even when, at 2 P.M., Bateman got word that "C" company of 1/9th Gurkhas held Hangman's Hill, he could not get any support to them as the Germans still held the strong positions between Castle and Hangman's hills.[155] Bateman ordered another attack on the points after dark. This time 1/6th Rajputana Rifles were able to storm Pt. 236, but could not hold it. They withdrew under counterattack just before dawn. The action had, however, allowed "D" Company of 1/9th Gurkhas to traverse a path past German defensive positions to reinforce "C" Company on Hangman's Hill.[156] Together they repulsed a strong German counterattack.[157]

The 4th Indian Division now held Hangman's Hill, a key point above the Monastery, but maintained only a fragile connection between the units there and the rest of the forces. According to the official British history, "To send supplies to Hangman's Hill was still more difficult because the porters had to cross a further thousand yards of bare no-mans-land, swept by fire and a fair ground for German fighting patrols. Needless to say, that movement was possible only during hours of darkness."[158] 4/6th Rajputana Rifles were ordered to act as porters to resupply the men on Hangman's Hill, but they could only carry so much per night.[159] Efforts were made to airdrop supplies to the position, but the terrain offered few decent landing spots and the supplies often broke up and spread down the mountain.[160]

In the meantime, 2nd New Zealand Division had achieved some success in Cassino town, enabling them to attack and take Pt. 202 from the east. However, at the same time the Germans reinforced their hold in some ruined houses below Castle Hill (possibly using tunnels from the monastery), threatening the 4th Indian Division position there. As the Indian official historian Dharm Pal, notes, "This pocket of German resistance added considerably to the difficulties of the 5th Indian Infantry Brigade, because the ground for the deployment of troops on Monastery Hill was now limited to a route, through the Castle itself."[161] General Freyberg hoped that control over Pt. 202, Castle Hill, and Hangman's Hill would be sufficient launching point for a final offensive against the monastery.[162] The plan was to move 1/4th Essex up to Hangman's Hill to join the 1/9th Gurkhas, while 4/6th and 1/6th Rajputana Rifles would be combined into a composite unit at Castle Hill. They would then launch an attack on the Monastery from both positions.

But on March 19, before these movements could be completed, the Germans counterattacked.[163] 4/6th Rajputana Rifles were overwhelmed, and the Germans proceeded past them to attack Castle Hill. By the morning, only three officers and 60 men were left in the Castle, but they fought off two waves of German assaults, demonstrating good discipline. By midnight a company of 2/7th Gurkhas arrived to reinforce the Castle, helping the smattering of soldiers from 1/4th Essex, 1/6th Rajputana Rifles, and 4/6th Rajputana Rifles to fend off the final German attack.[164] In the meantime, the companies from 1/4th Essex that had been sent to reinforce Hangman's Hill were caught on the edge of the battle; only 70 soldiers arrived at the position.[165]

This battle of March 19 destroyed Freyberg's hope of launching a strong attack on the Monastery. That night another series of moves was made to try to address the new German threat to Castle Hill, shifting a company of 4/6th Rajputana Rifles and one of 2/7th Gurkhas from the castle to Pt. 175, some distance north of the castle, now threatened by German reinforcements.[166] The next night, March 20, the 6th Royal West Kent regiment and 2/7th Gurkha Rifles were sent west to try to capture Pts. 236 and 445, both positions to the north of the monastery, in an effort to cut off German forces there from those around the castle and in Cassino town. Neither unit made much progress, the 2/7th Gurkhas making three attempts before abandoning the effort and returning to the castle.[167] At the same time, the Germans counterattacked the Rajputana Rifles and 2/7th Gurkhas at Pt. 175. The Indian soldiers succeeded in defending their positions and fighting off three German attacks over the course of the night.

By March 22, both sides of the battle were holding their positions, but neither had the strength to mount a serious counterattack. According to Pal, imperial forces decided to abandon any further offensives on March 23.[168] Indian soldiers from the 4th Indian Division as well as New Zealand troops from the 2nd New Zealand Division had fought off elite German paratroopers and had succeeded in capturing some important positions. However, they were unable to pry the Germans fully from their positions in Cassino town, the monastery, or on the mountain.

In my study of the official histories and archival record, I found no evidence of failed discipline among Indian units in Cassino. That said, I also found no evidence of initiative, and there were questions regarding morale. Units ordered to take on difficult assignments, like the 4/6th Rajputana Rifles who were ordered to act as porters to supply Hangman's Hill, completed their tasks, but there is no evidence that soldiers volunteered for such assignments or tried out new ways to attack or defend their objectives. Freyberg questioned the state of morale as he was forced to continuously put off the start date of the Third Battle, although censorship reports for this final week of fighting noted that "combatant troops show an excellent fighting spirit."[169] Additionally, there is some evidence of Indian soldiers growing frustrated with their British colleagues. One British officer recalled attempting to make his way down one of the narrow mountain tracks and happening on a group of Sikh soldiers. It is worth quoting the exchange in full:

> One Sikh informed him that he could choose between two tracks down the hill, one covered by snipers, the other shorter but horribly precipitous. "One way you get shot, Sahib, the other you slip to your death. But," he grinned without humor, "if you do go that way (pointing to the short cut) you are on your own, we will not move to help you, even if you are hit." "Why?" I asked. "Why? Why? The Sahib asks why? Because we've already lost men doing crazy things for British Sahibs. Now do them for yourself." The others nodded in agreement.[170]

While this may have been an isolated incident, it highlights the fact that Indian units were being asked by British officers to take intensely dangerous actions and saw themselves as distinct from those officers and were unmoved by the goals those tasks were meant to achieve.

ITALY—BRITISH AT ANZIO

Background

As noted above, the landings at Anzio were an effort to flank the German Winter Line and draw German forces away from Cassino. According to the official history, "Clark and Lucas [Officers commanding the Allied landing forces] expected that the landings would be strongly opposed, and that heavy counter-attacks were certain."[171] Yet in the early morning hours of January 22, 1944, there was no reaction to the pre-landing naval shelling on the beaches. Allied forces had achieved complete surprise.[172]

The 6th US Corps, commanded by American Major-General John Lucas (6th Corps was part of the US 5th Army, commanded by Lt. General Mark Clark) initially consisted of two American divisions (3rd Infantry and 1st US armored Division) as well as the British 1st Infantry Division and 2nd Special Service Brigade. The British 56th Division would reinforce

6th Corps in early and mid-February. The surprise achieved in the initial landing allowed 6th Corps to quickly capture Anzio port, solving potential logistical issues and enabling a quick landing of forces originally intended to be held in reserve. However, Clark and Lucas were concerned about German counteroffensives and proceeded with caution. We will pick up the campaign with the initial Allied effort to break out of their beachhead, beginning on January 24, 1944.

Perceptions of the War

British perceptions of the war against Germany in Italy were essentially the same as they had been in North Africa. The British believed the war with Germany was compatible with several aspects of their national identity: It was a war to defend Britain and Europe from tyranny and to protect democracy, to protect the principle of self-determination that Hitler was violating in Europe, and to protect the British homeland. As the war moved into Italy, the same goals applied. The fact that the Italians had changed sides in September 1943 increased British contempt for their military but also helped to frame the fight in Italy as one to free the Italians from German oppression, something British troops witnessed firsthand.[173]

The British public supported the invasion of the continent and had a generally positive though realistic attitude toward the conduct of the war and the potential for victory. In polling at the end of 1943, 66 percent believed it was necessary for the Allies to invade the continent in order to defeat Germany.[174] Seventy-four percent indicated that they were satisfied with the government's conduct of the war to that point.[175] By January 1944 that number had fallen to 69 percent, likely due to the so-called March to Rome stalling at Cassino.[176] When asked how long they believed the war would last, only 18 percent thought it would be over within six months. Interestingly for this specific campaign, while 45 percent of those polled stated that they had feelings of hatred, bitterness, and anger toward the German people, only 5 percent felt that way toward the Italian people. The largest category, at 20 percent, felt sorry for them.[177]

The sense that victory was likely, even if it would take some time, meant that discussions of plans for postwar Britain were beginning. Polling indicated that the British anticipated a return to party politics; 41 percent indicated that they wanted the coalition government to continue after the war, while 46 percent wanted a return to party governance.[178] With this future shift in mind, politicians were beginning to float plans for postwar housing, health care, and social security programs for veterans.

Being on the European continent seemed to make soldiers feel that they were getting closer to the finish line, and thus home. One soldier wrote, "The lads out here have a slogan, 'Blighty or Bust," and they mean it. We're just 'homers' and we're marching home through Italy, Germany and France and any other blooming place they want to put in our path."[179] Another wrote, "I am now seeing something nearer to civilization and it is having its effect on me. This taste of Italy is making me more homesick every day.... We are ready for Civvy Street once more with the well-founded notion that we have done our bit."[180] Soldiers who had spent the last two to three years in the desert were closing in on victory and home, and were beginning to think about it in practical terms.

Though Italy did remind soldiers more of home than the Libyan desert had, British racism influenced the way soldiers viewed Italians. Unlike the Indian soldiers, who had generally positive opinions about the civilians around them, British soldiers looked down on them as both cowards and racially less-than. Although some initially felt sorry for Italian poverty, "Further acquaintance usually results in a considerable modification of this favourable first impression, however, and the 'old hands' usually dislike and despise the Italians, whom they describe as being dirty, lazy, treacherous and avaricious—the phrase 'little better than Wogs' is often used."[181] Though some, especially in the countryside, reported positive experiences with civilians, censors suggested that most soldiers felt contempt toward them and complained of being taken advantage of by local traders. British racism influenced soldiers' perceptions of the people they believed they had liberated.

Importantly, British soldiers in Italy also indicated that they were paying attention to the goals of the war and to the political situation at home. In November 1943 Oswald Mosley, a leading British fascist who was interned at the beginning of the war, was released. With Mosley being hotly debated in Parliament, British soldiers in Italy took notice. One soldier, writing home in December 1943 said: "You ask me what I think of Mosley being free. Well, I'm speechless over the whole affair. It sometimes makes me wonder if we are really fighting Fascism or making the world fit for Fascists to live in. I look upon it as a stab in the back for us chaps out here who are trying to stamp out everything Mosely stands for."[182] Another soldier wrote: "I would give anything for a night's sleep, I have never had my clothes off for a month, it is just get down here and there for a few hours then up and on the go again. And then you happen to get a look at an old newspaper and you find they have gone and let Oswald Mosley loose, talk about getting disheartened."[183] The censors highlight the quote as "typical of the majority viewpoint regarding the subjects concerned."[184]

Numerous letters home make mention of the goals of the war—defeating Germany for good, ending fascism, bringing an end to German atrocities against civilians, as well as improving British postwar society.[185] One soldier, writing in March 1944, saw himself and his British compatriots as key to obtaining the goals of the war: "I only hope that the United Nations and democracy and the people realise just how much they owe the 'Micks' [Irish] and of course the Jocks [Scottish] and Grens [Grenadier Guards], who have borne the brunt of the German attacks and broken them."[186] This soldier included the Irish volunteers specifically along with the Scottish members of his unit, noting the multinational makeup of the unit defending democracy and the "United Nations."

Thinking about postwar life, another soldier was concerned about the seeming growth of state control over life, writing, "I wonder what Britain will offer after the war—will it be a mass-produced life of Jerry-built houses, communal eating, of utility clothes and furniture, and government inspectors, or will there be the pre-war family life, with a certain amount of personal choice and individual wishes?" Alternatively, another saw postwar planning as comforting: "I am extremely pleased to listen to the many discussions and plans being made for post-war Britain, it gives one a feeling that he is not fighting for nothing, and has his future welfare assured by the state."[187] Censors assessed the overall comments in letters thus: "Apart from the sceptics, who declare that little will be done for the future ex-soldier and that an ungrateful country will forget all about him, there appear to be two main viewpoints regarding post-war problems: That of the men who demand sweeping changes and drastic legislation; and that of those who are beginning to fear that this would result in a country dragooned and disciplined out of democracy, and the loss of the freedom they are fighting for."[188] British soldiers in Italy kept abreast of political debates at home and thought about those debates in terms of their relationship to the goals they believed themselves to be fighting to achieve in Italy.

It should be noted that race makes an appearance in discussions of postwar planning as well. One soldier wrote home, "Another big fly in the ointment is the hold that the Jews have on everything in Blighty, we think one of our first jobs will be to loosen that stranglehold a bit."[189] Despite witnessing German atrocities against Italy and information being available about the destruction of Europe's Jewish population, some British soldiers continued to view politics through antisemitic tropes.

The Campaign

After a nearly uncontested landing, the 6th Corps spent two days consolidating the beachhead and bringing forward supplies and reserves. On January 25, 1944, it began to move eastward from the beachhead, American forces on the right and British forces on the left. The 24th Guards Brigade, of the 1st British Infantry Division, was ordered to attack up one of the two main roads in the region, via Anziate, toward the town of Aprilia, which was held by a German panzer battalion.[190] The 5th Grenadier Guards Battalion, followed by the Irish Guards and Scotts Guards battalions, moved toward Aprilia and engaged the Germans who were occupying a factory compound outside the town. Made up of factory buildings as well as a model village for the workers, the area provided significant cover. The 5th Grenadiers' first company strength attack failed, but an attack by the full battalion supported by artillery fire was successful.[191] The 5th Grenadier Guards now occupied the factory buildings and were supported by the Irish guards occupying an unused railway bed just west of the factory area, with the Scots Guards in reserve.

The next morning the Germans counterattacked with an infantry battalion supported by tanks. The 5th Grenadier Guards and Irish Guards repulsed the attack, even destroying four tanks without themselves having armored support.[192] In the first engagement of the battle at Anzio, British forces demonstrated discipline in taking on German tanks without armored support. The Germans followed up with air attacks, but the 24th Guards Brigade retained the position.

During the week between the landings and the first breakout effort, German General Kesselring was able to bring in significant reinforcements. Though the Allies had hoped for just such an action, they had anticipated that Kesselring would have to withdraw forces from Cassino to reinforce Anzio, thus easing the way for the second battle of Monte Cassino, which was intended to begin in mid-February. Instead, replacement forces from Germany were rushed to Italy.[193] When the 6th Corps attempted to break out of the beachhead at Anzio, it faced significant German forces who were themselves organizing for a counterattack, but without having drawn forces away from Cassino.

The 6th Corps planned to attack on a two-division front—US 3rd Division on the right and British 1st Division on the left with US 1st Armored Division. The chosen route for US 1st Armored Division was what the official historian referred to as a "tank trap."[194] It contained gullies fifty feet deep and muddy bogs that trapped the tanks; eventually the division would ask for and receive permission to use the main road up which the British had attacked.[195] Yet the effort on the left was to be the main effort in seeking to break out of the beachhead. Launching this attack also required the Scots and Irish Guards Battalions to seize the launching point for the division just south of the town of Campoleone, northeast of Aprilia. The original plan had been for the Grenadier Guards to lead the attack, but the day before (January 28) eight officers of the battalion got lost returning from a scouting mission and drove straight into a German outpost. Three were killed in the escape and one seriously injured, leaving the battalion missing a significant number of its officer corps. While new officers were put in place, the Scots and Irish Guards Battalions replaced the 5th Grenadier Guards in the attack on Campoleone.[196]

At 11 p.m. on January 29, the Scots Guards launched an attack on the east-west road south of Campoleone and immediately struck a mined and wired roadblock. Fighting through significant defense, the Scots Guards were able to seize their objectives but with heavy losses. The Irish Guards, attacking the road to the left of the Scots Guards, hit even stronger defenses and obtained their objective only at dawn. First light, however, exposed them to German tanks, forcing them to withdraw, in order, south of the Scots Guards' position.[197] The Irish Guards' objective required the combined efforts of the King's Shropshire Light Infantry (KSLI) of the 3rd Infantry Brigade and the tanks of the 46th Royal Tank Regiment. While the Irish Guards were unable to overcome dug-in German tanks, they did fight through substantial obstacles and when they withdrew, did so in an organized manner to the agreed position, demonstrating discipline.

Having captured the road south of Campoleone, the Irish and Scots Guards battalions now held the position, while the 3rd Infantry Brigade and the 46th Royal Tank Regiment launched the planned (but delayed) attack on Campoleone. The 1st Shropshire and the 1st Duke of Wellingtons seized their objectives, the high ground around the railway station, by 5:30 p.m. when further action was called off.[198]

The 3rd Infantry renewed its advance the next day (January 31) with the 2nd Foresters supported by the 46th Royal Tank Regiment. Both units ran into heavy resistance from German forces dug into the houses around the railway, and steep embankments prevented the British tanks from proceeding to support the infantry. After several hours stalled in this situation, artillery support was called in to "soften up" the area, but even then, the Foresters and 46th Royal Tank Regiment were stopped by heavy fire from houses that had essentially become a series of German tank and machinegun emplacements.

At this point the US 1st Armored Division had abandoned their attempt at moving up the flank and instead followed the British forces up the main road. Major General Harmon, commander of the US forces, moved to relieve a group of British Foresters holding a position on the high ground above Campoleone. He later wrote that he had never "seen so many dead men in one place."[199] Of 116 men, only 16 remained, the highest ranking of whom was a corporal. Harmon quotes him as saying, "We're ordered to hold out until sundown, and I think, with a little good fortune, we can manage to do so."[200] Despite losing their officers and most of their comrades, the men of the Foresters were willing to remain in place and hold their position against strong German fire, demonstrating strong discipline. Nevertheless, the Foresters and supporting tanks were withdrawn, leaving the Shropshires and the Duke of Wellingtons holding the high ground just south of the railway.[201]

Because the American efforts to the right of the British had also stalled out in the face of significant German resistance, the result of the British efforts was to create a salient into the German defenses. The 3rd Brigade was ordered into defensive positions and reinforced with the 2nd Infantry Brigade. Because intelligence reports indicated that the Germans were preparing a counterattack in force, the Allied forces shifted from offensive to defensive orientation.

The counterattack began on February 3 as the British 168th Brigade arrived to reinforce the 1st Infantry Division. (The rest of its home division, the 56th Infantry division, would arrive in mid-February.) The British-held salient would be the location of the most severe battles over the coming month, made more difficult by the fact that it was full of topographical tank traps—ridges, wadis, and embankments that made tank support of infantry very difficult.[202]

The German effort began with small attacks against two battalions of the 3rd Brigade at the apex of the salient—the 1st Duke of Wellington's regiment and the KSLI, which they easily repulsed.[203] The main thrust was directed against the flanks of the salient, in the rear of these two units. German panzer regiments infiltrated the positions of the 6th Gordons on the east, splitting the battalion in half, while the German 65th Infantry did the same to the Irish Guards in the west.[204] Demonstrating initiative, the Irish Guards organized a counterattack to prevent the Germans from completely cutting the 3rd Infantry Brigade off from the rear.[205] The 6th Gordons, reinforced with a battalion from the fresh 168th Brigade and two squadrons of tanks, retook the ground they had lost that morning. That allowed the KSLI and the 1st Duke of Wellington to be withdrawn to more defensible positions that evening.[206] That night General Lucas visited the 3rd Brigade, noting in his diary that he found the British "all right," with morale elevated due to the recovery of some members of the Duke of Wellington believed to have been lost.[207]

On February 7 the Germans renewed their attack on the salient, focusing their attention on the North Staffordshires (recently attached to the 24th Guards Brigade) and 5th Grenadier Guards who held the Buonriposo ridge overlooking Aprilia. Both units took serious losses, fighting small hand-to-hand engagements in slit trenches and scrub.[208] Demonstrating initiative, a small group of 5th Grenadier Guards held a bridge over a gulley, slowing tanks by throwing hand grenades. One officer was wounded and a soldier killed, but the Germans withdrew.[209] The North Staffordshires fared worse and were slowly pushed off the ridge. Showing strong discipline, however, they continued to engage the enemy until they ran out of ammunition.[210] At the end of the battle, the North Staffordshire had lost seventeen officers and 364 men. The regiment was reorganized into a single rifle company.

Meanwhile on February 7 the Germans also attacked the factory area at Aprilia, held by the 168th Brigade. Fierce small-unit engagements broke out as the Germans infiltrated among the factory buildings.[211] Two companies of the 10th Royal Berkshires (168th Brigade) were overrun, but they continued to fight and engage in counterattacks throughout the next two days.[212] Despite some reinforcements and counterattacks on the flanks, the British lost control of Aprilia and withdrew south under orders.[213]

During the night of February 9–10 the 1st Division reorganized itself and prepared defenses along its new line, from just south of Buonriposo ridge, covering the town of Carroceto, and now south east Aprilia.[214] The following night the Germans moved south from the ridge, attacking the now isolated Scots Guards and Irish Guards who held the railway station at Carroceto.[215] Again the battle was a series of isolated company-sized engagements. The Guards counterattacked repeatedly, and some units engaged in hand-to-hand combat. But by morning the British could no longer hold the station, and the Guards fell back to the 1st Division's main line to the south.[216]

After a failed American attempt to retake the factory area at Aprilia, major fighting died down to patrolling while both sides reorganized and brought up reinforcements. The British 56th Division arrived, and the 168th Brigade rejoined its home division, which relieved the 1st Division on the main defensive line. It now shared that line with the US 45th Division to the right. The Germans attacked again on February 16.

Based on reports from prisoners that the Germans were planning a counterattack on February 16, Allied artillery and air forces dropped more than 700 tons of bombs on the German positions.²¹⁷ The Germans were not deterred and launched their attack on the 56th British Infantry Division and US 45th Infantry Division as planned.²¹⁸ Both were able to repel the initial attack, but it was renewed that night, primarily against US forces.

On February 17 the British 1st Division occupied the Allied forces' final defense lines behind the US 45th Division, while the 179th Infantry (US) counterattacked against the Germans. The Germans held and on February 18 took advantage of US disorganization to attack through all the way to the British 1st Division line. According to the official history, "the main weight of the attack fell on the [1st] Loyals [of 2nd Infantry Brigade] who gave no ground and restored the local situation by immediate counterattack."²¹⁹ The Germans then turned their attention to the remains of the 179th [US] Infantry, which had taken up position to the left of the Loyals, and then again attacked the Loyals from the right.²²⁰ Coordinating with the North Staffordshires and the 46th Royal Tank Regiment, the Loyals counterattacked the Germans and pushed them back from the final defensive line.²²¹ British units, including those that had been in battle for several weeks with little rest, demonstrated discipline and initiative, staying in the fight and seeking engagement with the enemy where possible. Though there were some small German efforts in the area over the next few days, most of the major fighting was done by February 20. A sort of trench warfare developed in the area over the next few months, but there were no large operations until May when Allied forces broke the stalemate.²²²

In addition to the positive assessment of morale noted by some American officers above, British censorship reports indicate that morale was good among British units in Italy throughout this period. The reports were based on letters from soldiers throughout Italy, not just those at Anzio, so they should be read with caution. That said, letters from the first two weeks of March, just after the main actions in Anzio were concluded, contained "evidence of a very high level of morale. Forward troops, in particular, while frequently admitting that 'the going is tough,' show a splendid fighting spirit, and every confidence."²²³ Two weeks later the censors write, "Many troops refer soberly to the 'tough going' in Italy, and not a few describe themselves as being considerably 'shaken' by their experiences, particularly in the Anzio bridgehead." However, they continue, "all troops appear to share the strong conviction that the defeat of Germany must be completed in thorough fashion," although they did hope that Russia would be a big part of that defeat.²²⁴

Pride in their unit appears in several letters home, with one soldier writing: "The Battalion is still going strong and keeping up its good name. . . . They're hot."²²⁵ Another, writing of a unit that was definitely present in Anzio, noted:

"Our men have fought hard and have made a good name for themselves and the Regt. I am sure it will go down in history, in fact, that the Shropshires fought such a fight on the Anzio beach head and I am very proud to be just a little cog in such a fighting machine as the Shropshires."[226] Some sense of esprit de corps seems to have been motivating soldiers in these units.

The official history suggests that a dip in morale was due in part to a perceived lack of effort to break out of the beachhead and that a change in command improved the situation at the end of February.[227] Overall, British soldiers demonstrated discipline and initiative throughout the battles at Anzio, with a potentially slight shake in morale due to a perceived lack of offensive action. They showed high will to fight under very difficult circumstances.

CONCLUSION

All three militaries in these cases faced difficult terrain, weather, and well-trained German opponents. Australian and British soldiers demonstrated high levels of will, showing initiative, discipline, and strong morale. Indian units had adequate will to fight; they showed good discipline, but morale varied and they did not demonstrate initiative in battle.

National identity theory, combined again with small-unit cohesion in the form of esprit de corps, explains will to fight well in these cases. British and Australian soldiers saw themselves as fighting to defend democracy and to defeat tyranny. Australians also saw their fight in Greece as playing an important part in the defense of the empire of which they were a part. British soldiers were also proud of their units and discussed their successes both in terms of the honor it brought to the unit and as contributing to victory over Nazism.

While cohesion played a role in Indian and British will, it was not sufficient to drive high will to fight among the Indian soldiers who saw no link between the goals of the battle and their national identities. Although Indian soldiers wrote of their pride in the successes and reputation of the 4th Indian Division, it was not sufficient to spark initiative. It needed to be linked to national identity to spark the emotion necessary for initiative. Neither was cohesion necessary for the discipline, morale, and initiative demonstrated by Australian soldiers in Greece.

Lack of democracy at home was clearly a source of frustration among Indian soldiers, and there was political repression in India during this period. Yet democracy did not do the expected work among Australian and British soldiers, with soldiers of both nationalities strongly criticizing domestic politicians and indicating that they felt a lack of support from the home front. Level of threat to the homeland was a factor in British will, but it was minor and did not play the expected role at all among Australian and Indian soldiers. Rather, soldiers interpreted threat through their sense of identity. Australian soldiers believed they were protecting Australia by defending the empire. Indian soldiers saw the threat to their families as coming from the famine and failure of the British to follow through on promises rather than the Japanese.

NOTES

1. Churchill, *His Complete Speeches*, 6100.
2. Cruikshank, *Greece 1940–1941*, 53.
3. Playfair, *Volume II*, 83.
4. As quoted in Long, *Greece, Crete, and Syria*, 15.
5. Hill, *Diggers and Greeks*, 12.
6. As quoted in Long, 17.
7. As quoted in Long, 19.
8. As quoted in Hasluck, *The Government and the People 1939–1941*, 336.
9. Cruikshank, 105; Playfair, *Volume II*, 24.
10. Long, 17.
11. AWM54 534/5/24 Part 1.
12. Long, 20.
13. Buckley, *Greece and Crete 1941*, 29.
14. Buckley, 43; Long, 57–58.
15. AWM 54 534/5/24 Part 1.
16. Hill, 28.
17. Churchill, 6372.
18. Churchill, 6372.
19. Hill, 27.
20. Hill, 50.
21. Robertson, *Australia at War 1939–1945*, 40.
22. National Archives of Australia (hereafter NAA) CP 313/1/20, emphasis mine.
23. Elkin, *Our Opinions*, 11.
24. Elkin, 14, 17.
25. NAA CP 313/1/21.
26. AWM 54 534/5/24 Part 1; AWM 54 534/2/37.
27. Hill, 70.
28. Johnston, *Fighting the Enemy*, 32.
29. TNA CAB 106/555.
30. TNA CAB 106/555.
31. Johnston, *Fighting the Enemy*, 33.
32. As quoted in Hill, 129.
33. TNA CAB 106/555.

34. Johnston, *Anzacs in the Middle East*, 77.
35. Long, 46.
36. Buckley, 45; Long, 47.
37. Long, 47, 53, 56.
38. Long, 47.
39. Long, 57.
40. Buckley, 56.
41. Long, 57.
42. AWM 54 534/5/14.
43. Buckley, 55.
44. As quoted in Hill, 89.
45. Long, 60.
46. Long, 60.
47. Long, 61.
48. AWM 54 534/2/37.
49. Hill, 89.
50. Buckley, 55.
51. Buckley, 55.
52. Long, 65 (footnote 4).
53. Long, 63.
54. Long, 64–65.
55. As quoted in Long, 67.
56. Long, 57.
57. Long, 81.
58. Long, 72.
59. AWM 54 534/5/24; Long, 107.
60. Long, 113–14.
61. AWM 54 534/5/24 Part 1.
62. Long, 115.
63. Buckley, 87.
64. Long, map between pp. 106–7.
65. Long, 116.
66. Long, 118.
67. AWM54 534/5/24 Part 1.
68. AWM 54 534 5/24 Part 1.
69. AWM 54 534/5/24 Part 1.
70. AWM 54 534/2/35 Part 2.
71. AWM 54 534/2/35 Part 2.
72. AWM 54 534/2/35 Part 2.
73. As quoted in Long, 143.
74. Long, 146.
75. Long, 146.
76. Long, 147.
77. Long, 143.
78. Long, 159.
79. AWM 54 534/2/35 Part 2.
80. Long, 166.

81. AWM 54 534/2/35 Part 2.
82. AWM 54 534/2/35 Part 2.
83. Long, 167.
84. Long, 168.
85. Long, 168.
86. AWM 54 534/2/35 Part 2.
87. Long, 180.
88. Long, 181.
89. AWM 54 534/2/35 Part 2.
90. AWM 54 534/2/35 Part 2.
91. Long, 182–83.
92. NAA CP 313/1 21.
93. NAA CP 313/1 21.
94. Ellis, *Cassino*, 68.
95. Molony, *Volume V*, 691.
96. Ellis, 114
97. Pal, *The Campaign in Italy*, 91.
98. Ellis, 161.
99. Owen, "The Cripps Mission," 80.
100. Owen, 81.
101. Owen, 81.
102. TNA WO 106/3720.
103. Khan, *India at War*, 134.
104. TNA WO 106/3720.
105. Khan, 147.
106. TNA WO 106/3720.
107. BL IOR L/PJ/12/655.
108. BL IOR L/PJ/12/655.
109. TNA WO 204/10381.
110. TNA WO 106/3720; Khan, 180.
111. Khan, 183. (Italics in original.)
112. TNA WO 106/3720.
113. See, for example, Mukherjee, *Hungry Bengal*.
114. Mukherjee, 178.
115. Mukherjee, 191.
116. Mukherjee, 191.
117. Mukherjee, 175.
118. TNA WO 204/10381.
119. BL IOR L/PJ/12/655; Khan, 190.
120. BL IOR L/PJ/12/655.
121. BL IOR L/PJ/12/655.
122. BL IOR L/PJ/12/655.
123. BL IOR L/PJ/12/655. Translated by censors.
124. BL IOR L/PJ/12/655.
125. TNA WO 106/3720.
126. TNA WO 106/3720.
127. BL IOR L/PJ/12/655.

128. BL IOR L/PJ/12/655.
129. TNA WO 204/10381.
130. Ellis, 175.
131. Ellis, 176–77.
132. Ellis, 179.
133. TNA WO 204/10381.
134. Ellis, 183.
135. Molony, 717.
136. Pal, 114.
137. Molony, 717.
138. Pal, 112.
139. Pal, 113.
140. Pal, 113.
141. Molony, 717.
142. Pal, 113.
143. Ellis, 214.
144. Pal, 119.
145. Molony, 783–84.
146. Pal, 125.
147. Ellis, 223.
148. Molony, 780.
149. Molony, 788.
150. Molony, 788.
151. Pal, 129.
152. Pal, 130.
153. Molony, 788; Pal, 130.
154. Molony, 789.
155. Molony, 790.
156. Molony, 791.
157. Pal, 130.
158. Molony, 794.
159. Pal, 131.
160. Molony, 795.
161. Pal, 133.
162. Pal, 133.
163. Pal, 134; Molony, 797.
164. Pal, 135.
165. Molony, 797.
166. Pal, 138.
167. Pal, 139.
168. Pal, 141.
169. TNA WO 204/10381.
170. As quoted in Ellis, 255.
171. Molony, 658.
172. TNA CAB 106/413.
173. TNA WO 204/10381.
174. Gallup, *The Gallup International Public Opinion Polls*, 84.

175. Gallup, 84.
176. Gallup, 88.
177. Gallup, 82.
178. Gallup, 83.
179. TNA WO 204/10381.
180. TNA WO 204/10381.
181. TNA WO 204/10381.
182. TNA WO 204/10381.
183. TNA WO 204/10381.
184. TNA WO 204/10381.
185. TNA WO 204/10381.
186. TNA WO 204/10381.
187. TNA WO 204/10381.
188. TNA WO 204/10381.
189. TNA WO 204/10381.
190. Molony, 669; TNA CAB 106/413.
191. TNA CAB 106/413.
192. TNA CAB 106/413.
193. Molony, 661.
194. Molony, 672.
195. Eisenhower, *They Fought at Anzio*, 140.
196. Molony, 675.
197. TNA CAB 106/413.
198. TNA CAB 106/413.
199. As quoted in Eisenhower, 141.
200. Eisenhower, 141.
201. TNA CAB 106/413.
202. Molony, 727.
203. Molony, 728.
204. Molony, 728.
205. Molony, 729; TNA CAB 106/413.
206. Molony, 729.
207. As quoted in Eisenhower, 157.
208. TNA CAB 106/413.
209. Molony, 733.
210. TNA CAB 106/413.
211. TNA CAB 106/413.
212. Molony, 734.
213. Molony, 734; CAB 106/413.
214. TNA CAB 106/413.
215. Molony, 735.
216. Molony, 735; TNA CAB 106/413.
217. Molony, 745.
218. TNA CAB 106/413.
219. Molony, 748.
220. TNA CAB 106/413.
221. TNA CAB 106/413.

222. Molony, 757.
223. TNA WO 204/10381.
224. TNA WO 204/10381.
225. TNA WO 204/10381. (emphasis in original).
226. TNA WO 204/10381.
227. Molony, 751–52.

CONCLUSION

The previous chapters argued that national identity theory can explain variation in will to fight among units of different nationalities fighting under the auspices of the British Empire during WWII. The case studies demonstrate substantial evidence for my claims, explaining most of the variation among British, Indian, and Australian soldiers in campaigns in North Africa, Malaya, and Europe. When soldiers identify the cause of their fight with their sense of national identity, it leads to high levels of will to fight—to morale, discipline, and initiative in battle.

This final chapter takes up two tasks. First, while this book has focused on a particular imperial state in a particular war, national identity theory should help to explain will to fight in other groups and conflicts. I briefly consider a case from a different time and nation to determine the plausibility of national identity theory as an explanation beyond World War II—the Union Army in the US Civil War. Second, I consider the theoretical and policy implications of my findings and consider directions for future research.

BEYOND WORLD WAR II?

Table C.1 summarizes the findings of in chapters 3, 4, and 5. National identity theory (NIT) can explain all cases except that of Indian units' high levels of will to fight in North Africa. Cohesion at the organizational level is better able to explain Indian motivation in North Africa, but, as noted below, the evidence is not able to fully test small-unit cohesion (SUC) as a whole. What it does suggest is that commitment to the organization can at times be compounded by commitment to other forms of identity, but at times the two can conflict. National identity also informs soldiers' understanding of what constitutes a threat to their nation. It is not simply geographic proximity to an opponent but an opponent that threatens the myths and symbols of their self-identity. Furthermore, the cases bear out the accuracy of the theory's causal logic. National identity theory can explain the process by which units developed their will to fight in all but one case.

Table C.1: Theoretical Explanations for Levels of Will to Fight

	North Africa	Malaya	Europe
British	High—NIT	Adequate—NIT	High—NIT
Indian	High—SUC	Poor/Adequate—NIT/SUC	Adequate—NIT/SUC
Australian	High—NIT	High—NIT	High—NIT

This book has focused on a single, though large and complex, conflict. Comparing three nationalities all fighting the same opponent under very similar material conditions enabled me to hold constant a number of important factors and focus on the role that national identity and the cause of the war played in motivating—or not motivating—soldiers in battle. I have offered significant evidence that British, Australian, and Indian soldiers considered the goals of the war in which they fought in the context of the myths and symbols of their national identities. Those soldiers who found compatibility fought with more will than those who did not. But this approach to evaluating the usefulness of my argument does not demonstrate how generalizable it is. Could my argument explain will to fight in soldiers of other nationalities or in wars besides WWII? A brief look at the US Civil War suggests that national identity theory could plausibly travel beyond the British Empire in World War II.

US Civil War

The American Civil War was a war about slavery: For the South, the right to own slaves and to extend slavery into the West; for the North, it began as a war to preserve the Union in spite of slavery and ended as a war to end slavery in order to preserve the Union. I focus on the Union side of the conflict, although there is significant evidence that the Confederate side was also motivated by the belief that the cause was fundamental to preserving their identity. While both sides referenced some of the same myths and symbols—republicanism, liberty, the Minutemen, and Founding Fathers—there was significant disagreement regarding whether these myths and symbols were compatible with secession. In some sense, the Civil War was in part about the fact that there was no single national identity within the United States at the time.[1] Even on the Union side, there was disagreement over emancipation as a war goal and slavery's relationship to American identity.

Yet many Union soldiers were clearly motivated by their belief that the fight to preserve the Union was not only compatible with but necessitated by their national identity; it was fundamental to preserving liberty, order, and America as a "city on a hill."

James McPherson, Chandra Manning, and John Hess, among other Civil War historians, argue that significant numbers of Union soldiers were motivated by their belief that the goals of the war were compatible with and even necessitated by their sense of national identity. In his study of the letters and diaries of 647 Union soldiers, McPherson finds that 68 percent indicated patriotism as a motivation after their enlistment.[2] Manning and Hess support this claim. All three find significant evidence of soldiers drawing on the myths and symbols of national identity to explain their willingness to fight. References to the legacy and sacrifice of 1776, the preservation of democracy, and (white) political liberty are present throughout the writings of Union soldiers.[3] In particular, many soldiers called on the founding myth of the United States as a "city on a hill," an example of liberty and democratic governance to the people of Europe ruled by monarchs and other tyrants. Secession threatened to confirm European doubts about democracy and to harm the cause of liberty around the world.[4]

Abolition and emancipation were more contentious goals than preservation of the Union. Early on in the war there were a number of soldiers who believed that emancipation was necessary, although their reasoning varied. For some, it was a moral opposition to holding humans as property.[5] For many, it was simply the fact that the South was seceding to preserve slavery so slavery must be removed as the root cause of disunion (though others used this same logic to argue for accepting slavery to preserve the Union).[6] But many Union soldiers, especially those from the border states, wanted to preserve both slavery and the Union.[7] McPherson argues that this segment of the population was part of the reason that Lincoln initially hesitated to make the war one of emancipation. Indeed, when the Emancipation Proclamation was made public, it provoked significant debate among soldiers as to whether it was an appropriate goal. While a minority continued to oppose emancipation throughout the war, by 1863 many Union soldiers had embraced it.[8] For some this was due to exposure to the brutality of slavery during their time in the South.[9] For others, a pragmatic attitude dominated: the enslaved contributed to the South's war economy, while Black Union soldiers contributed to the Union Army's strength. More free Black men would mean less labor for the South and more soldiers for the North.[10] Whatever their reasons, enough soldiers were comfortable with emancipation as a war goal that 80 percent of Union soldiers voted for Lincoln over the anti-emancipation candidate (and their former commander), George McClellan.[11]

We can draw some broad conclusions about how belief in the cause as compatible with their identity motivated soldiers in battle. McPherson found that soldiers with more ideological commitment to the war were more likely to have volunteered and to reenlist when their original enlistment period was up. He even presents evidence of volunteering for difficult assignments.[12] Soldiers who believed in the cause of the war were more likely to show initiative.

The Union Army was not a particularly well-disciplined organization as a whole. The prewar Army had significant desertion rates, which did not improve on expansion for the war.[13] Additionally, Union soldiers were truly committed democrats; they elected their own officers and objected to efforts to impose on their individualism.[14] But commitment to the war does seem to have been linked to discipline. In a representative sample of 31,854 white enlisted men, Dora Costa and Matthew Kahn find that ideology was an important predictor of desertion and straggling. Men who enlisted early in the war or who came from counties that voted for Lincoln were less likely to desert or be arrested.[15]

Morale appears to be linked both to belief in the cause and to the progress of the war. The lowest points of Union Army morale come at moments of defeat, particularly the summer of 1864.[16] Additionally, soldiers noted with deep frustration the growing support for the peace democrats back home.[17] Much like the Australians in North Africa and the British in Italy, Union soldiers were frustrated by a lack of commitment to the war effort back home.

A sense of national identity and the belief that the goals of the war were compatible with that national identity appear to have played a role in Union soldiers' will to fight during the Civil War. Soldiers were deeply engaged with the politics of the war, discussed it in terms of national myths and symbols, and debated whether specific goals were compatible with those myths and symbols. The available evidence suggests that there was at least some connection between a belief that the war was compatible with their national identity and soldiers' willingness to fight.

The detailed studies of British, Indian, and Australian combat motivation in WWII, as well as this brief look at Union soldiers in the US Civil War, suggest that national identity theory is a plausible explanation for combat motivation. Further research is necessary to determine just how generalizable that explanation is.

THEORETICAL IMPLICATIONS

This book makes three important theoretical contributions to the study of international security. First, it provides a deeper understanding of combat effectiveness and ultimately the conduct of war. National identity theory contends that the political debate around a war and its goals or purpose matters to soldiers and ultimately to the conduct of the war. Soldiers in each of the nationalities examined here paid close attention to the reasons leaders gave for sending them to war as well as to ongoing domestic political debates. British soldiers approved of the goals of the war against Germany—defending the British homeland as well as freedom and democracy in Europe. They compared those goals with the goal of defending the empire from Japanese aggression and found the war against Japan less important. Australian soldiers were committed to the goals of both wars but were consistently frustrated by domestic politics, which they saw as undermining the war efforts. Indian soldiers understood the goals of the war and were unconvinced that fighting to preserve the British Empire was worth their sacrifice. This was especially the case when the British rulers of India reneged on their promise to take care of soldiers' families at home when famine and disease took hold. Soldiers are not apolitical machines but are engaged with the reasons they fight and the politics of the leaders who send them into battle.

The cases in this book have interesting implications for small-unit cohesion's contribution to combat effectiveness. It is important to note that, given my focus on the brigade and battalion levels, the cases do not test arguments pertaining to platoon-level cohesion. There are some instances of companies (100–200 men) being combined successfully, but the available archival evidence does not reach to the platoon level that authors like Shils and Janowitz argue is key to motivation. Nevertheless, the cases are suggestive.

First, the cases above indicate that some forms of secondary-group cohesion (esprit de corps, unit honor, organizational loyalty) can improve will to fight. Indian soldiers who fought in North Africa were committed to the "regiment" (usually the battalion or even division) that they joined, to its history and connection to their home communities. They were inspired by the honor and history of the larger group of which they were a part. Indian soldiers who joined newer units in the expanded Indian Army or were from communities without a tradition of army service do not appear to have had a strong sense of identification with the organization.

Secondary-group cohesion was not always sufficient to sustain motivation. Sikh soldiers were willing to directly disobey their officers, risking courts-martial and dishonor to the unit when orders threatened their religious commitments. Members of the 4th Indian Division who expressed pride in their division's accomplishments against Rommel in North Africa became increasingly politicized and frustrated with the war efforts as they witnessed India's suffering in famine from thousands of miles away in Italy. And the British Argylls in Malaya, who had a strong sense of esprit de corps and were led by an experienced and respected officer in Col. Ian Stewart, still found themselves with only adequate will to fight when they did not understand or believe in the goals of the war. Secondary-unit cohesion mattered, but other identities (religious or national) could compete.

Understanding the sources of soldiers' will to fight is a key element in understanding combat effectiveness and sources of military power more broadly. Material capabilities like weaponry and force size clearly matter in the outcomes of the battles in this book. The British imperial forces lacked airpower and anti-tank weaponry in Malaya, for example, putting them at a severe disadvantage when facing Japanese tanks. Both German and British imperial forces were less successful as they got further from their base of supplies. And the geography of Italy limited the kinds of tactics available to British and Indian forces at Monte Cassino and Anzio, despite their significant numbers and firepower. Will alone cannot explain success or failure in battle. Yet the cases also demonstrate that it clearly matters. Australian soldiers were able to overcome some of Japan's superiority in armored vehicles by abandoning fixed defensive positions and adopting new tactics and new uses for the weapons they did have. If British and Indian soldiers had also taken the initiative in battle with the Japanese, they may not have defeated that materially superior force, but they may have imposed higher costs in the course of the battle, shifting the long-term balance of forces in the Pacific. Soldiers with high will to fight will put their equipment and training to better use than those with low levels of will.

One limit of the research design is that it is unable to determine exactly how much combat motivation mattered relative to other important elements of effectiveness such as technology, doctrine, or economic resources. For example, significant evidence both in this book and other scholarship suggests that soldiers' battlefield skills are an important element of combat effectiveness.[18] Yet I cannot determine through my case studies whether motivation could overcome lack of skills. Moreover, it is possible that militaries can overcome lack of motivation, given sufficient resources and political will. For example, Jason Lyall argues that Soviet units were effective in some of their battles despite a lack of combat motivation because of blocking units—units who would shoot soldiers who retreated from battle.[19] Given the Soviet advantage in manpower, motivation was not always necessary for victory.

The second theoretical contribution is that national identity theory draws together important literatures on political myth, emotion, and nationalism to offer a clear causal logic for the operation of nationalism in combat. Nationalism is not simply a response to threat or the result of rallying around the flag at the onset of any conflict. Rather, national identity works to increase will to fight when soldiers believe that the goals for which they are asked to fight are compatible with the myths and symbols of their national identity. Leaders are constrained by existing myths and symbols as well as by the facts of the war. Soldiers assess compatibility between the facts of the war as they know them and the myths and symbols of their national identity. If they believe the goals of the war to contradict their identity, they will be less willing to fight that war, showing lower levels of morale, discipline, and initiative in battle. Elites play a role in mobilizing a population for war by calling on the myths and symbols of national identity, but they are constrained in two vital ways: they cannot create new myths, and lying about the goals of the war is risky and difficult in the long term.

Additionally, national identity does not need to include xenophobia or racism in order to inspire will to fight. Some national identities include racism against outgroups. But not all national identities include such myths, and neither is it necessary to dehumanize the enemy in order to motivate soldiers. The British, at both the elite and popular levels, demonstrated racist attitudes toward the Japanese, yet did not demonstrate high will to fight in Malaya. On the other hand, neither of the groups held racist attitudes toward the Germans in North Africa. In fact, both groups indicated a sense of professional respect for their opponent. Yet Australian and British soldiers demonstrated high will to fight against the Germans at the same time. There is some evidence that racist or xenophobic beliefs about opponents may lead to brutality against them in war zones, but it does not appear that soldiers must dehumanize their opponent in order to be willing to fight against them.[20]

Third, the argument and evidence in this book have implications for understanding the role of emotion and identity in international relations more broadly. If national identity theory is correct, it furthers our understanding of the nature of military power. Material analysis alone is not sufficient for assessing a state's potential threat or level of security. While technology, military size, and economic wealth all matter, so too does a state's ability to translate that material capability into battlefield performance. That ability is limited and shaped by the relationship between the myths of a political community's identity and the purpose to which a state's leaders want to turn its material capability.

The 2022 Russian invasion of Ukraine provides an important case in point. Western governments did not have high expectations for the Ukrainian military. Despite improvements in training and equipment since 2014, some Western governments believed that Ukraine would likely be overwhelmed by a full-scale invasion. Russia was simply bigger, better equipped, and better trained.[21] In part these analysts were correct—the Russian military is bigger, generally better equipped, and—on paper—better trained. But the Ukrainian military had something that the Russian military did not: commitment to the goals of the war.[22] The Russian forces were in some cases unaware that they were even going to war, believing themselves to be on exercises. In other cases, they were shocked to find that rather than welcoming them as liberators, Ukrainian civilians fought back however they could.[23] And Ukrainians soldiers are committed to defending their homeland from invasion by a country of which they used to be a part.

Additionally, national identity theory argues that behavior among both leaders and publics is influenced by emotion, especially emotion sparked by reference to myths and symbols of identity. My focus in this book has been narrow: how does identity spark emotion under conditions of war, and how does that emotion influence behavior on the battlefield? But my findings suggest that emotion may be more fundamentally important in international relations than many of our theories currently acknowledge. The findings here suggest that the study of international relations needs to take more seriously the contributions of constructivist and feminist scholars who have been challenging rationalist and materialist assumptions in the discipline for some time.

Future Research

There are two clear areas for future research based on the argument and findings in this book. First, my cases all come from large-scale conventional war. While the evidence suggests national identity theory is a strong explanation for will to fight in such conflicts, does the argument capture will in smaller or unconventional wars? Does cause matter to members of small all-volunteer forces or during counterinsurgency operations? Does it capture will to fight in nonstate militia groups or only in military organizations associated with states? Further research is necessary to explore the generalizability of the argument.

Second, I have focused on the role of national identity in explaining soldiers' will to fight. But the causal logic may extend beyond national identity to other forms of identity as well. Evidence from the Indian Army cases suggests that religious identity may play a similar role to national identity in shaping will during war. However, religion in those cases was closely tied to national identities, so we should approach those findings with caution. A group such as ISIS would provide a better case study for the influence of religious identity on will to fight, as ISIS explicitly rejected the nation-state ideal type as part of their ideology. While national identity and religious identity appear to be the most common forms of political identification in current international politics, history offers alternative forms of communal identification that may operate in similar ways. More work is necessary to understand whether all forms of communal or political identification can serve as a source of motivation in combat, or whether there is something particular to national identity that makes it unique.

POLICY IMPLICATIONS

Finally, national identity theory and the findings in this book have important implications for policymakers. As noted above, soldiers' identities and their relationship with the goals of particular wars shape a military's overall effectiveness. This should be accounted for in force assessment efforts. The argument does not provide a neat formula but does suggest important factors that should be added to overall force assessment efforts: the level of national identification within a state, the content of its national identity, and elite rhetoric around particular points of conflict are all important for assessing a military's potential will and thus effectiveness.

Second, the argument suggests that much of the debate regarding manpower policies in modern militaries is misplaced. There is no need to worry about social cohesion within small units as it is neither necessary nor sufficient for creating will to fight. Debates about including ethnic and racial minorities, women, and LGBTQ+ individuals in the military have relied heavily on claims about the impact integrating such "others" might have on small-unit cohesion. National identity theory suggests that interpersonal relationships are not as fundamental to will to fight as Shils and Janowitz, among others, claim.[24] Rather, leaders should instead focus on ensuring their soldiers are clear on the goals for which they fight and that those goals line up with their sense of common identity.

It is here that national identity theory does suggest limits on manpower policy. A sense of common belonging is important for motivating soldiers to fight. As noted earlier, some national identities are more inclusive than others. Exclusive national identities would limit the number of people who identify with them and thus limit the number of people who would be motivated by calls to the myths and symbols of those identities. There is growing concern in the United States and elsewhere that racially or sexually exclusive national identities are a growing threat not only to political equality but to national security. Jason Lyall argues that militaries marked by inequality, often linked with exclusive national identities, are less effective and successful in war.[25] National identity theory complements this claim by demonstrating that national identity is a key source of will to fight. If that national identity excludes many people within the population, it limits the number of potential soldiers who would be motivated to fight for that identity. In this way national identity theory suggests that in the long term states that are able to develop more inclusive national identities and more equal societies will be more secure in the international system.

Finally, national identity theory sounds a note of caution about the decision to use force. Leaders cannot simply assume that soldiers will be motivated to fight for any ends the leader pursues through the use of military force. Soldiers are not particularly amenable to propaganda. They pay attention to politics and consider policies in light of deeply held identities. Elites should consider whether the myths and symbols held by their populations can sustain a commitment to a particular fight. As Krebs and Walldorf have argued, national narratives constrain policymakers by limiting the options behind which they can rally popular support.[26] The findings in this book suggest that constraint is not simply political but also military; leaders are constrained by the narratives for which their soldiers are willing to fight.

NOTES

1. Hess, *Liberty*, 28.
2. McPherson, *For Cause and Country*, 101.
3. McPherson, 18; Manning, "A Vexed Question," 33; Hess, 19.
4. Manning, 33; McPherson, 112.
5. Manning, 48.
6. Manning, 34.
7. McPherson, 121.
8. McPherson, 123.
9. Manning, 34.

10. McPherson, 125.
11. Manning, 48.
12. McPherson, 81; 115.
13. Reid and White, "Stragglers and Cowards," 66.
14. Reid and White, 68.
15. Costa and Kahn, "Cowards and Heroes," 539.
16. Costa and Kahn, 539; McPherson, 146.
17. McPherson, 144.
18. See especially chapter 4 discussion of the 12th Indian Infantry Brigade and Talmadge, *Dictator's Army*.
19. Jason Lyall, "Forced to Fight."
20. Bartov, "The Conduct of War;" Browning, *Ordinary Men*; Dower, *War Without Mercy*.
21. Heinrich and Sabes, "Gen. Milley,"
22. Onuch and Hale, *The Zelensky Effect*.
23. The Guardian, "Demoralised Russian Soldiers," March 4, 2022.
24. Shils and Janowitz, "Cohesion and Disintegration;" Simmons, "Bad Idea."
25. Lyall, *Divided Armies*.
26. Krebs, *Narrative*; Walldorf, *To Shape Our World for Good*.

BIBLIOGRAPHY

PRIMARY SOURCES

British National Archives (TNA)
https://www.nationalarchives.gov.uk/
- CAB 106/53
- CAB 106/54
- CAB 106/55
- CAB 106/56
- CAB 106/162
- CAB 106/413
- CAB 106/555
- WO 106/2550A
- WO 106/2550B
- WO 106/3720
- WO 169/3351
- WO 169/3443
- WO 204/10381
- WO 208/673

British Library (BL)
https://www.bl.uk/
- IOR L/PJ/12/654
- IOR L/PJ/12/655

Australian War Memorial (AWM)
https://www.awm.gov.au/
- 52 8/3/18/12
- 54 534/2/35 Part 2
- 54 534/2/37
- 54 534/5/14
- 54 534/5/24 Part 1
- 54 553/1/6
- 54 553/5/1
- 54 883/2/97 Part 1
- 54 883/2/97 Part 2
- SALT Volume 1
- SALT Volume 2

National Archives of Australia (NAA)
https://www.naa.gov.au/
   CP313/1/20
   CP313/1/21

SECONDARY SOURCES

Ahmed, Sara. *The Cultural Politics of Emotion*. Edinburgh University Press, 2014.

Ambrose, Stephen E. *Citizen Soldiers: The U.S. Army from the Normandy Beaches to the Bulge to the Surrender of Germany, June 7, 1944–May 7, 1945*. Simon and Schuster, 1997.

Anderson, Benedict. *Imagined Communities: Reflections on the Origin and Spread of Nationalism*. Verso, 1991.

Archer, Margaret. "Introduction: Realism in the Social Sciences." In *Critical Realism: Essential Readings*, edited by Margaret Archer, Roy Bhaskar, Andrew Collier, Tony Lawson, and Alan Norrie, pp. 189–205. Routledge, 2013.

Ariffin, Yohan, Jean-Marc Coicaud, and Vesselin Popovski, eds. *Emotions in International Politics: Beyond Mainstream International Relations*. Cambridge University Press, 2016.

Atkinson, Rick. *The Day of Battle: The War in Sicily and Italy, 1943–1944*. Henry Holt Company, 2007.

Atlantic Charter, August 14, 1941. https://avalon.law.yale.edu/wwii/atlantic.asp

Atran, Scott, Hammad Sheikh, and Angel Gomez. "Devoted Actors Sacrifice for Close Comrades and Sacred Cause." Proceedings of the National Academy of Sciences, 111, no. 50 (2014): 17702–17703.

Balzacq, Thierry. "A Theory of Securitization: Origins, Core Assumptions, and Variants." In *Securitization Theory: How Security Problems Emerge and Dissolve*, edited by Thierry Balzacq, pp. 1–30. Taylor and Francis, 2010.

Bar, Niall. *Pendulum of War: The Three Battles of El Alamein*. Overlook Press, 2005.

Barkawi, Tarak. *Soldiers of Empire: Indian and British Armies in World War II*. Cambridge University Press, 2017.

Bartov, Omer. "The Conduct of War: Soldiers and the Barbarization of Warfare." *The Journal of Modern History* 64, Supplement; *Resistance Against the Third Reich* (December 1992): S32–S45.

Bartov, Omer. *The Eastern Front, 1941–1945: German Troops and the Barbarisation of Warfare*. Palgrave, 2001.

Barua, Pradeep P. *Gentlemen of the Raj: The Indian Army Officer Corps, 1817–1949*. Praeger, 2003.

Bashford, Alison, and Stuart Macintyre, eds. *The Cambridge History of Australia* 4. Cambridge University Press, 2013.

Baum, Matthew, and Philip B. K. Potter. "The Relationship Between Mass Media, Public Opinion, and Foreign Policy: Toward a Theoretical Synthesis." *Annual Review of Political Science* 11 (2008): 39–65.

Bayly, Christopher, and Tim Harper. *Forgotten Armies: The Fall of British Asia, 1941–1945*, Belknap Press of Harvard University Press, 2005.
Baynes, John. *Morale: A Study of Men and Courage*. Cassell, 1967.
BBC. WW2 People's War, https://www.bbc.co.uk/history/ww2peopleswar/timeline/factfiles/nonflash/a6651218.shtml.
Ben-Shalom, Uzi, Zeev Lehrer, and Eyal Ben-Ari. "Cohesion During Military Operations." *Armed Forces and Society* 32, no. 1 (2005): 63–79.
Bhargava, K.D. *Official History of the Indian Armed Forces in the Second World War 1939–1945: Campaigns in South-East Asia 1941–1942*. Edited by Bishewar Prasad. Combined Interservices Historical Section, 1960.
Bharucha, P.C. *Indian Armed Forces in World War II: The North African Campaign 1940–1943*. Edited by Bisheshwar Prasad. Combined Inter-Services Historical Section, 1956.
Bhattacharya, Sanjoy. *Propaganda and Information in Eastern India 1939–1945: A Necessary Weapon of War*. Curzon, 2001.
Biddle, Stephen. *Military Power: Explaining Victory and Defeat in Modern Battle*. Princeton University Press, 2004.
Biddle, Stephen, and Stephen Long. "Democracy and Military Effectiveness: A Deeper Look." *The Journal of Conflict Resolution* 48, no. 4 (2004): 525–546.
Bierman, John, and Colin Smith. *The Battle of Alamein: Turning Point, World War II*. Viking, 2002.
Black, Jeremy. *English Nationalism: A Short History*. C. Hurst, 2018.
Bleiker, Roland, and Emma Hutchison, "Fear No More: Emotions and World Politics." *Review of International Studies* 34, no. S1(2008): 115–135.
Bloom, William. *Personal Identity, National Identity and International Relations*. Cambridge University Press, 1990.
Bond, Brian. *British Military Policy Between the Two World Wars*. Clarendon Press, 1980.
Boren, Cindy. "Colin Kaepernick Protest Has 49ers Fans Burning Their Jerseys." *The Washington Post*, August 28, 2016. https://www.washingtonpost.com/news/early-lead/wp/2016/08/28/colin-kaepernick-protest-has-49ers-fans-burning-their-jerseys/.
Bottici, Chiara, and Angela Kuhner. "Between Psychoanalysis and Political Philosophy: Towards a Critical Theory of Political Myth." *Critical Horizons* 13, no. 1, (2012): 94–112.
Brathwaite, Kirstin J. H. "Boys Will Be Boys? The Normative Sources of Prostitution Policy in the German and American Militaries During World War II." *Journal of Global Security Studies* 2, no. 1 (January 2017): 39–54.
Brathwaite, Kirstin J. H. "Effective in Battle: 'Conceptualizing Soldiers' Combat Effectiveness." *Defence Studies* 18, no. 1 (2018): 1–18.
Brathwaite, Kirstin J. H., and Margarita Konaev. "War in the City: Ethnic Geography and Combat Effectiveness." *Journal of Strategic Studies* (October 2019): 33–68.

Brooks, Risa. "Introduction: The Impact of Culture, Society, Institutions, and International Forces on Military Effectiveness." In *Creating Military Power*, edited by Risa Brooks and Elizabeth Stanley. Stanford University Press, 2007.

Browning, Christopher. *Ordinary Men: Reserve Police Battalion 101 and the Final Solution in Poland*. Harper, 2017.

Bruscino, Thomas. "The Analogue of Work: Memory and Motivation for Second World War US Soldiers." *War and Society* 28, no. 2 (2009): 85–103.

Buckley, Christopher. *Greece and Crete 1941*. H. M. Stationery Office, 1977.

Burke, Gerald, and Andrew Spaull. "Australian Schools: Participation and Funding 1901 to 2000." In *Year Book Australia 2001*, 1301. Australian Bureau of Statistics. https://www.abs.gov.au/ausstats/abs@.nsf/Lookup/A75909A2108CECAACA2569DE002539FB.

Cantril, Henry, ed. *Public Opinion 1935–1946*. Princeton University Press, 1951.

Carissimo, Justin. "People Are Burning Colin Kaepernick Jerseys over His Refusal to Stand During the National Anthem." *The Independent*, August 28, 2016. https://www.independent.co.uk/news/people/people-are-burning-colin-kaepernick-jerseys-over-his-refusal-to-stand-during-the-national-anthem-a7214281.html.

Carver, Michael. *Dilemmas of the Desert War: A New Look at the Libyan Campaign*. Batsford in Association with the Imperial War Museum, 1986.

Castillo, Jasen. *Endurance and War*. Stanford University Press, 2014.

Castinera, Angel. "Imagined Nations: Personal Identity, National Identity, and the Places of Memory." In *Contemporary Majority Nationalism*, edited by Alain-G. Gagnon, Genevieve Nootens, and Andre Lecours, pp. 43–77. McGill-Queen's University Press, 2011.

Cederman, Lars Erik, T. Camber Warren, Didier Sornette. "Testing Clausewitz," *International Organizations* 65, no. 4, (2011).

Chacho, Tania. "The Influence of Ideology: Soldier Motivation and American Combat Infantrymen in the European Theater of Operations During the Second World War." (PhD Dissertation, Johns Hopkins University, 2005).

Chacho, Tania. "Why Did They Fight? American Airborne Units in World War II," *Defense Studies* 1, no. 3 (2001): 59–94.

Chatterjee, Partha. *The Nation and Its Fragments: Colonial and Postcolonial Histories*, Princeton University Press, 1993, ebook.

Chaudhary, Latika. "Caste, Religion and Fragmented Societies: Education in British India." LSE Blog, May 1, 2013, https://blogs.lse.ac.uk/southasia/2013/05/01/caste-religion-and-fragmented-societies-education-in-british-india/.

Churchill, Winston. *Blood, Sweat, and Tears*. G.P. Putnam's Sons, 1941.

Churchill, Winston. *Winston Churchill: His Complete Speeches 1897–1963, 6, 1935–1942*. Edited by Robert James. Chelsea House, 1974.

Cochran, Shawn. "The Civil-Military Divide in Protracted Small War: An Alternative View of Military Leadership Preferences and War Termination." *Armed Forces and Society* 40, no. 1 (2014): 71–95.

Cohen, Stephen P. *The Indian Army: Its Contribution to the Development of the Nation.* University of California Press, 1971.
Colley, Linda. *Britons: Forging the Nation 1707–1837.* Yale University Press, 2005.
Connable, Ben, Michael J. McNerney, William Marcellino, Aaron Frank, Henry Hargrove, Marek N. Posard, S. Rebecca Zimmerman, Natasha Lander, Jasen J. Castillo, and James Sladden. *Will to Fight: Analyzing, Modeling, and Simulating the Will to Fight of Military Units.* Rand Corporation, 2018.
Costa, Dora, and Matthew Kahn, "Cowards and Heroes: Group Loyalty in the American Civil War." *The Quarterly Journal of Economics* 118, no. 2 (May 2003): 519–548.
Crawford, Neta. "Institutionalizing Passion in World Politics: Fear and Empathy." *International Theory* 6, no. 3 (2014): 535–557.
Crawford, Neta. "The Passion of World Politics: Propositions on Emotion and Emotional Relationships." *International Security* 24, no. 4 (2000): 116–156.
Cruickshank, Charles. *Greece 1940–41.* Davis-Poynter, 1976.
Darwin, John. *The Empire Project: The Rise and Fall of the British World-System 1830–1970,* Cambridge University Press, 2009.
Desch, Michael. *Power and Military Effectiveness: The Fallacy of Democratic Triumphalism.* Johns Hopkins University Press, 2008.
Doherty, Richard. *Irish Men and Women in the Second World War.* Four Courts Press, 2021.
Donnel, John C., Guy J. Pauker, and Joseph J. Zasloff. *Viet Cong Motivation and Morale in 1964: A Preliminary Report.* Rand, 1965.
Dower, John. *War Without Mercy: Race and Power in the Pacific War.* Pantheon Books, 1993.
Eddy, John, and Deryck Schreuder. *The Rise of Colonial Nationalism: Australia, New Zealand, Canada and South Africa First Assert Their Nationalities, 1880–1914.*, Allen & Unwin, 1988.
Edelman, Murray. *The Symbolic Uses of Politics.* University of Illinois Press, 1967.
Edwardes, Michael. *British India: 1772–1947.* Sidgwick and Jackson, 1967.
Einwohner, Rachel. "Opportunity, Honor, and Action in the Warsaw Ghetto Uprising of 1943." *American Journal of Sociology* 109, no. 3 (November 2003), 650–75.
Eisenhower, John S.D. *They Fought at Anzio.* Columbia University Press, 2007.
Elkin, A. P. *Our Opinions and the National Effort.* Australasian Medical, 1941.
Ellis, John. *Cassino: The Hollow Victory.* Andre Deutsch, 1984.
Esch, Joanne. "Legitimizing the 'War on Terror': Political Myth in Official-Level Rhetoric." *Political Psychology* 31, no. 3 (2010): 357–391.
Farrell, Brian P. *The Defence and Fall of Singapore 1940–1942.* Tempus Publishing, 2006.
Fennell, Jonathan. "Courage and Cowardice in the North African Campaign: The Eighth Army and Defeat in the Summer of 1942." *War in History* 1, no. 2 (2013): 99–122.
Fennell, Jonathan, *Fighting the People's War: The British and Commonwealth Armies and the Second World War.* Cambridge University Press, 2019.
Ferguson, Niall. *The Pity of War.* Basic Books, 1999.

Flood, Christopher G. *Political Myth: A Theoretical Introduction*. Routledge, 2002.
French, David. "Discipline and the Death Penalty in the British Army in the War Against Germany During the Second World War." *Journal of Contemporary History* 33, no. 4 (1998): 531–545.
French, David. *Military Identities: The Regimental System, the British Army, and the British People, 1870–2000*. Oxford University Press, 2005.
French, David. *Raising Churchill's Army: The British Army and the War Against Germany, 1919–1945*. Oxford University Press, 2001.
French, David. "'You Cannot Hate the Bastard Who is Trying to Kill You...': Combat and Ideology in the British Army in the War Against Germany, 1939–1945." *Twentieth Century British History* 11, no. 1 (2000): 1–22.
Fritz, Stephen. *Frontsoldaten*. University Press of Kentucky, 1995.
Fuller, J.G. *Troop Morale and Popular Culture in the British and Dominion Armies 1914–1918*. Oxford University Press, 1990.
Fussell, Paul. *The Great War and Modern Memory*. Oxford University Press, 2013.
Gal, Reuven, and Fredrick Manning. "Morale and Its Components: A Cross-National Comparison." *Journal of Applied Social Psychology* 17, no. 4 (1987): 369–391.
Gallup, George G., ed. *The Gallup International Public Opinion Polls: Great Britain 1937–1975*. Random House, 1976.
Gandhi, M.K. *Hind Swaraj or Indian Home Rule*. Navajivan Publishing, 1938. https://www.jmu.edu/gandhicenter/_files/gandhiana-hindswaraj.pdf.
Gelpi, Christopher, and Peter Feaver. "Iraq the Vote: Retrospective and Prospective Foreign Policy Judgments on Candidate Choice and Casualty Tolerance." *Political Behavior* 29, no. 2 (2007): 151–174.
Ginges, Jeremy, and Scott Atran, "War as a Moral Imperative (Not Just Practical Politics by Other Means)." *Proceedings of the Royal Society* (B 2782011): 2930–2938.
Gould, William. *Hindu Nationalism and the Language of Politics in Late Colonial India*. Cambridge University Press, 2004.
Grainger, J.H. *Patriotisms: Britain 1900–1939*. Routledge & Kegan Paul, 1986.
Grant, Lachlan. *Australian Soldiers in Asia-Pacific in World War II*. New South Publishing, University of New South Wales, 2014.
Grauer, Ryan. "Why Do Soldiers Give Up? A Self-Preservation Theory of Surrender." *Security Studies* 23 (2014): 622–655.
Grauer, Ryan, and Stephen L. Quackenbush. "Initiative and Military Effectiveness: Evidence from the Yom Kippur War." *Journal of Global Security Studies* (2021): 1–18.
Greenwood, Arthur (MP). *Why We Fight: Labour's Case*. Routledge, 1940.
Grey, Jeffrey. *A Military History of Australia*. Cambridge University Press, 2008.
The Guardian. "Demoralised Russian Soldiers Tell of Anger at Being 'Duped' into War," March 4, 2022. https://www.theguardian.com/world/2022/mar/04/russian-soldiers-ukraine-anger-duped-into-war.
Hall, Todd. "We Will Not Swallow This Bitter Fruit: Theorizing a Diplomacy of Anger." *Security Studies*. 20(2011): 521–555.

Hannah-Jones, Nikole, et al. "The 1619 Project." *The New York Times Magazine*, April 14, 2019. https://www.nytimes.com/interactive/2019/08/14/magazine/1619-america-slavery.html.

Hasan, Mushirul, ed. *Towards Freedom: Documents on the Movement for Independence in India, 1939*, 1. Oxford University Press, 2008.

Hasluck, Paul. *The Government and the People 1939–1941*. Australian War Memorial, 1952.

Hastings, Max. *Inferno: The World at War 1939–1945*. Alfred Knopf, 2011.

Hauser, William L. "The Will to Fight." In *Combat Effectiveness: Cohesion, Stress, and the Volunteer Military*, edited by Sam C. Sarkesian, 185–211. Sage, 1980.

Hayes, Jarrod. "Identity, Authority, and the British War in Iraq." *Foreign Policy Analysis* 12, no. 3 (July 2016): 334–353.

Heinrich, Jacqui, and Adam Sabes. "Gen. Milley Says Kyiv Could Fall Within 72 Hours if Russia Decides to Invade Ukraine." Fox News, February 5, 2022, https://www.foxnews.com/us/gen-milley-says-kyiv-could-fall-within-72-hours-if-russia-decides-to-invade-ukraine-sources.

Henderson, William Darryl. *Cohesion: The Human Element in Combat*. National Defense University Press, 1985.

Henderson, William Darryl. *Why the Vietcong Fought: A Story of Motivation and Control in a Modern Army in Combat*. Greenwood Press, 1979.

Hess, Earl. *Liberty, Virtue, and Progress: Northerners and Their War for the Union*. Oxford University Press, 1997.

Hill, Maria. *Diggers and Greeks*, University of New South Wales Press, 2010.

Hirst, John. "Empire, State, and Nation." In *Australia's Empire*, edited by Deryck Schreuder and Stuart Ward. Oxford University Press, 2008.

Horne, Gerald. *Race War! White Supremacy and the Japanese Attack on the British Empire*. New York University Press, 2004.

Hutchinson, John. *Nationalism and War*. Oxford University Press, 2017.

Imy, Kate. *Faithful Fighters: Identity and Power in the British Indian Army*. Stanford University Press, 2019.

Jachovic, Juljan. "Reinforcing the National Identity: Belarusian Identity-Building Social Practices." *Journal on Baltic Security* 8, no. 2 (2022): 3–41.

Jackson, Ashley. *The British Empire and the Second World War*. Hambledon Continuum, 2007.

Jackson, W.G.F. *The Battle for North Africa 1940–1943*. Mason/Charter, 1975.

Johnson, Dominic, and Dominic Tierney. *Failing to Win: Perceptions of Victory and Defeat in International Politics*. Harvard University Press, 2006.

Johnston, Mark. *Anzacs in the Middle East: Australian Soldiers, Their Allies, and the Local People in World War II*. Cambridge University Press, 2013.

Johnston, Mark. *Fighting the Enemy: Australian Soldiers and Their Adversaries in World War II*. Cambridge University Press, 2000.

Johnston, Mark. "The Civilians Who Joined Up, 1939–1945." https://www.awm.gov.au/articles/journal/j29/civils.

Kaufman, Stuart. *Modern Hatreds*. Cornell University Press, 2003.
Kaufman, Stuart. *Nationalist Passions*. Cornell University Press, 2015.
Kaura, Uma. *Muslims and Indian Nationalism: The Emergence of the Demand for India's Partition 1928–1940*. South Asia Books, 1977.
Kellett, Anthony, *Combat Motivation: The Behavior of Soldiers in Battle*. Kluwer, 1982.
Kershaw, Ian. "How Effective Was Nazi Propaganda?" In *Nazi Propaganda*, edited by David Welch, pp. 180–205. Croom Helm, 1983.
Khan, Yasmin. *India at War*. Oxford University Press, 2015.
Kier, Elizabeth. "Homosexuals in the U.S. Military: Open Integration and Combat Effectiveness." *International Security* 23, no. 2 (1998): 5–39.
King, Anthony "On Combat Effectiveness in the Infantry Platoon: Beyond the Primary Group Thesis." *Security Studies* 25, no. 4 (2016): 699–728.
King, Anthony. *The Combat Soldier: Infantry Tactics and Cohesion in the Twentieth and Twenty-First Centuries*, Oxford University Press, 2013.
King, Anthony, and Patrick Bury. "A Profession of Love: Cohesion in a British Platoon in Afghanistan." In *Frontline*, edited by Anthony King, pp. 200–215. Oxford University Press, 2015.
Kirby, Stanley Woodburn. *Singapore: The Chain of Disaster*. Cassell, 1971.
Kirby, Stanley Woodburn. *The War Against Japan: Volume 1*. M. Stationary Office, 1957.
Kirke, Xander. "Violence and Political Myth; Radicalizing Believers in the Pages of Inspire Magazine." *International Political Sociology* 9, no. 4 (2015): 283–298.
Kocher, Matthew, Adria Lawrence, and Nuno Monteiro. "Nationalism, Collaboration, and Resistance: France Under Nazi Occupation." *International Security* 43, no. 2 (2018): 117–150.
Krebs, Ronald. *Narrative and the Making of US National Security*. Cambridge University Press, 2015.
Kumar, Krishan. *The Making of English National Identity*. Cambridge University Press, 2003.
Kurki, Milja. "Critical Realism and Causal Analysis in International Relations." *Millennium: Journal of International Studies* 35, no. 2 (2007): 361–278.
LaCasse, Alexander. "'Divisive Concepts' Ban Is New Hampshire Law. Will It Affect the Way Teachers Do Their Jobs?" *Portsmouth Herald*, July 9, 2021, https://www.seacoastonline.com/story/news/local/2021/07/10/new-hampshire-education-divisive-concepts-ban-nh-law-affects-schools/7915398002/.
Lake, David. "Powerful Pacifists: Democratic States and War." *American Political Science Review* 86, no. 1 (1992): 24–37.
Lappin, Yaakov. "I Was Willing to Die to Stop the Syrian Advance." *The Jerusalem Post*, September 21 2015. https://www.jpost.com/Arab-Israeli-Conflict/I-was-willing-to-die-to-stop-the-Syrian-advance-416734.
Lehmann, Todd C., and Yuri M. Zhukov. "Until the Bitter End? The Diffusion of Surrender Across Battles." *International Organization* 73, no. 1 (2019): 133–69.
Lemmon, Gayle Tzemach. *The Daughters of Kobani*. Penguin, 2021.

Lepre, George. *Fragging: Why U.S. Soldiers Assaulted Their Officers in Vietnam*. Texas Tech University Press, 2011.

Lloyd, Amy J. "Education, Literacy and the Reading Public." *British Library Newspapers*. Gale, 2007.

Long, Gavin. *Australia in the War of 1939–1945: Series One, Volume II: Greece, Grete, and Syria*. Australian War Memorial, 1953.

Losh, Jack. "Britain's Abandoned Black Soldiers." *Foreign Policy*, https://foreignpolicy.com/2019/02/23/britains-abandoned-black-soldiers/.

Lowenthal, David. "The Island Garden: English Landscapes in British Identity." In *History, Nationhood and the Question of Britain*, edited by Helen Brocklehurst and Robert Philips. Palgrave MacMillan, 2004.

Lunn, Ken, and Ann Day. "Britain as Island: National Identity and the Sea." In *History, Nationhood, and the Question of Briton*, edited by Helen Brocklehurst and Robert Philips. Palgrave MacMillan, 2004.

Lyall, Jason. *Divided Armies*. Princeton University Press, 2020.

Lyall, Jason. "Forced to Fight: Coercion, Blocking Detachments, and Tradeoffs in Military Effectiveness." In *The Sword's Other Edge: Trade-offs in the Pursuit of Military Effectiveness*, edited by Dani Reiter. Cambridge University Press, 2017.

Lynn, John. *The Bayonets of the Republic: Motivation and Tactics in the Army of Revolutionary France, 1791–1794*. University of Illinois Press, 1984.

Macintyre, Stuart. *A Concise History of Australia*. Cambridge University Press, 1999.

MacKenzie, Megan. *Beyond the Band of Brothers: The US Military and the Myth That Women Can't Fight*. Cambridge University Press, 2015.

Maclean, Kama. *A Revolutionary History of Interwar India: Violence, Image, Voice, and Text*. Oxford University Press, 2015.

Malesevic, Sinisa. "The Structural Origins of Social Cohesion: The Dynamics of Micro-Solidarity in 1991–1995 Wars of Yugoslav Succession." *Small Wars and Insurgencies* 29, no. 4 (2018): 735–753.

Manning, Chandra. "'A Vexed Question:' White Union Soldiers on Slavery and Race." In *The View from the Ground: Experiences of Civil War Soldiers*, edited by Aaron Sheehan-Dean. University Press of Kentucky, 2007.

Markwica, Robin. *Emotional Choices: How the Logic of Affect Shapes Coercive Diplomacy*. Oxford University Press, 2018.

Marshall, Jonathan. *To Have and Have Not: Southeast Asian Raw Materials and the Origins of the Pacific War*. University of California Press, 1995.

Marshall, S.L.A. *Men Against Fire*. Combat Force Press. 1947, 1954.

Maughan, Barton. *Australia in the War of 1939–1945: Tobruk and El Alamein*. Australian War Memorial, 1965.

McCarthy, Caitlin. "The 'Lessons' of the United Daughters of the Confederacy Still Have Influence Today in the Mid-South." Local Memphis ABC. https://www.localmemphis.com/article/news/local/the-lessons-of-the-united-daughters-of-the-confederacy-still-have-influence-today-in-the-mid-south/522-aa3185da-142b-48b2-b64e-f44fa70e5309.

McDonnell, Michael A. *Masters of Empire: Great Lakes Indians and the Making of America.* Hill and Wang, 2015.
McKernan, M., and M. Browne, eds. *Australia: Two Centuries of War and Peace.* Australian War Memorial, 1988.
McLachlan, Noel. *Waiting for the Revolution: A History of Australian Nationalism.* Penguin, 1989.
McLean, Denis. *The Prickly Pair: Making Nationalism in Australia and New Zealand.* University of Otago Press, 2003.
McMinn, W.G. *Nationalism and Federalism in Australia.* Oxford University Press, 1994.
McPherson, James. *For Cause and Comrades: Why Men Fought in the Civil War.* Oxford University Press, 1997.
Mearsheimer, John. *Why Leaders Lie.* Oxford University Press, 2011.
Mercer, Jonathan. "Feeling Like a State: Social Emotion and Identity." *International Theory* 6, no. 3 (2014): 515–535.
Metcalf, Barbara, and Thomas Metcalf. *A Concise History of Modern India.* Cambridge University Press, 2012.
Minault, Gail. *The Khilafat Movement: Religious Symbolism and Political Mobilization in India.* Columbia University Press, 1982.
Molony, Brigadier C.J.C., Captain F.C. Flynn, Major-General H.L. Davies, and Group Captain T.P. Gleave. *The Mediterranean and the Middle East, Volume V: The Campaign in Sicily 1943 and the Campaign in Italy 3rd September 1943 to 31st March 1944.* Her Majesty's Stationary Office, 1973.
Moreman, T.R. *The Jungle, the Japanese, and the British Commonwealth Armies at War 1941–1945.* Frank Cass, 2005.
Morrow, James. "The Institutional Features of Prisoners of War Treaties." *International Organization* 55, no. 4 (2001): 971–991.
Moskos, Charles. *Soldiers and Sociology.* United States Army Research Institute for the Behavioral and Social Sciences, 1988.
Moskos, Charles. *The American Enlisted Man.* Basic Books, 1970.
Moyer, H. Wayne, "Ideology and Military Systems." In *Combat Effectiveness: Cohesion, Stress, and the Volunteer Military,* edited by Sam C. Sarkesian, pp. 108–153. Sage, 1980.
Mukherjee, Janam. *Hungry Bengal: War, Famine, and the End of Empire.* Oxford University Press, 2015.
Murray, Williamson. "British Military Effectiveness in the Second World War." In *Military Effectiveness* 3 edited by Allan R. Millett and Williamson Murray. Allen and Unwin, 1988.
Neillands, Robin. *Eighth Army: The Triumphant Desert Army That Held the Axis at Bay from North Africa to the Alps, 1939–45.* Overlook Press, 2004.
New York Times. *Transcript of Talk by Reagan on South Africa and Apartheid,* July 23, 1986. https://www.nytimes.com/1986/07/23/world/transcript-of-talk-by-reagan-on-south-africa-and-apartheid.html.
Onuch, Olga, and Henry E. Hale. *The Zelensky Effect.* Oxford University Press, 2023.

Owen, Nicholas. "The Cripps Mission of 1942: A Reinterpretation." *The Journal of Imperial and Commonwealth History* 30, no. 1 (2008).

Pal, Dharm. *Official History of the Indian Armed Forces in the Second World War 1939–1945: The Campaign in Italy 1943–1945*, edited by Bisheshwar Prasad. Delhi: Combined Inter-Services Historical Section, 1960.

Palazzo, Albert. *The Australian Army: A History of Its Organization 1901–2001.* Oxford University Press, 2001.

Pawinski, Michal, and Georgina Chami. "Why They Fight? Reconsidering the Role of Motivation in Combat Environments." *Defence Studies* 19, no. 3 (2019): 297–317.

Pearlman, Wendy. "Emotions and the Microfoundations of the Arab Uprisings." *Perspectives on Politics* 11, no. 2 (June 2013).

Petersen, Roger. "Identity, Rationality, and Emotion in the Process of State Disintegration and Reconstruction." In *Constructivist Theories of Ethnic Conflict*, edited by Kanchan Chandra. Oxford University Press, 2012.

Perry, F.W. *The Commonwealth Armies: Manpower and Organization in Two World Wars*, Manchester University Press, 1988.

Pfau, Ann. *Miss Yourlovin: GIs, Gender, and Domesticity During World War II.* Columbia University Press, 2013.

Pitt, Barrie. *The Crucible of War: Western Desert 1941.* Jonathan Cape, 1980.

Playfair, Major General I.S.O. *The Mediterranean and Middle East, Vol. II: The Germans Come to the Help of Their Ally (1941).* Her Majesty's Stationary Office, 1956.

Playfair, Major General I.S.O. *The Mediterranean and Middle East, Vol. III (September 1941 to September 1942): British Fortunes Reach their Lowest Ebb.* Her Majesty's Stationary Office, 1960.

Polity Project. *Polity5: Political Regime Characteristics and Transitions, 1800–2018.* Center for Systemic Peace, 2020. http://www.systemicpeace.org/inscr/p5manualv2018.pdf.

Posen, Barry. "Nationalism, the Mass Army, and Military Power." *International Security* 18, no. 2 (1993): 80–124.

Powell, David. *Nationhood and Identity: The British State Since 1800.* I.B. Tauris, 2002.

President's Advisory 1776 Commission, *The 1776 Report*, January 2021. https://trumpwhitehouse.archives.gov/wp-content/uploads/2021/01/The-Presidents-Advisory-1776-Commission-Final-Report.pdf.

Preston, Andrew. *Sword of the Spirit, Shield of Faith.* Alfred Knopf, 2012.

Raghavan, Srinath. *India's War: World War II and the Making of Modern South Asia.* Basic Books, 2016.

Raghuvanshi, V. P. S. *Indian Nationalist Movement and Thought.* L.N. Agarwal, 1959.

Reese, Roger. "Surrender and Capture in the Winter War and Great Patriotic War: Which Was the Anomaly?" *Global War Studies* 8, no. 1 (2011): 87–98.

Reese, Roger. *Why Stalin's Soldiers Fought: The Red Army's Military Effectiveness in World War II.* University Press of Kansas, 2011.

Reid, Brian Holden, and John White. "'A Mob of Stragglers and Cowards': Desertion from the Union and Confederate Armies, 1861–1865." *Journal of Strategic Studies* 8, no. 1 (1985): 65–77.

Reiter, Dan, and Allan Stam. *Democracies at War*. Princeton University Press, 2002.

Robertson, John. *Australia at War 1939–1945*. William Heinemann, 1981.

Rose, Sonya. *Which People's War? National Identity and Citizenship in Briton, 1939–1945*. Oxford University Press, 2003.

Rosentiel, Tom. "Public Attitudes Toward the War in Iraq: 2003–2008." Pew Research Center, March 19, 2008. https://www.pewresearch.org/2008/03/19/public-attitudes-toward-the-war-in-iraq-20032008/.

Ross, Andrew. *Mixed Emotions: Beyond Fear and Hatred in International Conflict*. University of Chicago Press, 2014.

Roy, Kaushik. *Battle for Malaya: The Indian Army in Defeat, 1941–1942*. Indiana University Press, 2019.

Sarkar, Sumit. *Modern India, 1885–1947*. MacMillan, 1983.

Sarkesian, Sam C. "Combat Effectiveness." In *Combat Effectiveness: Cohesion, Stress, and the Volunteer Military*, edited by Sam C. Sarkesian, pp. 8–18. Sage, 1980.

Seal, Anil. *The Emergence of Indian Nationalism: Competition and Collaboration in the Nineteenth Century*. Cambridge University Press, 1968.

Shils, Edward, and Morris Janowitz. "Cohesion and Disintegration in the Wehrmacht in World War II." *The Public Opinion Quarterly* 12, no. 2 (1948): 280–315.

Siebold, Guy. "The Essence of Military Group Cohesion." *Armed Forces and Society* 33, no. 2 (January 2007).

Sil, Rudra, and Peter J. Katzenstein. "Analytic Eclecticism in the Study of World Politics: Reconfiguring Problems and Mechanisms Across Research Traditions." *Perspectives on Politics* 8, no. 2 (June 2010).

Simmons, Anna. "Here's Why Women in Combat Units Is a Bad Idea." *War on the Rocks*, https://warontherocks.com/2014/11/heres-why-women-in-combat-units-is-a-bad-idea/.

Simunovic, Pjer. "The Russian Military in Chechnya—A Case Study in Morale in War." *Journal of Slavic Military Studies* 11, no. 1 (1998): 63–95.

Smith, Anthony. *Ethno-Symbolism and Nationalism: A Cultural Approach*. Routledge, 2009.

Smith, Colin. *Singapore Burning: Heroism and Surrender in World War II*. Viking, 2005.

Stanley, Peter. "What 'Battle for Australia'?" https://www.abc.net.au/news/2008-09-03/32530.

Stern, Paul C. "Why Do People Sacrifice for Their Nations?" *Political Psychology* 16, no. 2 (June 1995): 217–235.

Stewart, Nora Kinzer. *Mates and Muchachos: Unit Cohesion in the Falklands/Malvinas War*. MacMillan, 1991.

Talmadge, Caitlin. *The Dictators Army: Battlefield Effectiveness in Authoritarian Regimes*. Cornell University Press, 2015.

Taylor, Kate. "'Burn the NFL': Americans Are Destroying Football Jerseys After Players Kneel in Protest During the National Anthem." *Business Insider*, September 25, 2017. https://www.businessinsider.com/nfl-fans-burn-football-jerseys-after-anthem-protest-2017-9.

Tibbitts, Craig. "Australians in the First Battle of El Alamein July 1942." *Sabretache* 45, no. 1 (March 2004): 5–20.

Trench, Charles C. *The Indian Army and the King's Enemies, 1900–1947.* Thames and Hudson, 1988.

Van Creveld, Martin. *Command in War.* Harvard University Press, 1985.

Van Evera, Stephen. "Hypotheses on Nationalism and War." *International Security* 18, no. 4 (1994): 5–39.

Van Ness, Justin, and Erika Summers-Effler. "Emotions in Social Movements." In *The Wiley Blackwell Companion to Social Movements*, 2nd ed., edited by David Snow, Sarah A. Soule, Hanspeter Kriesi, and Holly J. McCammon, pp. 411–428. John Wiley and Sons, 2019.

Vosler, Brittany. "'Making His Way to the Heart of India': British Missionaries, Indian Nationalism, and Religious Belonging in Post-World War I India." *British Scholar* 3, no. 1 (2010): 61–78.

Walldorf, William C. *Just Politics: Human Rights and the Foreign Policy of Great Powers.* Cornell University Press, 2008.

Walldorf, William C. *To Shape Our World for Good.* Cornell University Press, 2019.

Walsh, James Igoe. "Precision Weapons, Civilian Casualties, and Support for the Use of Force." *Political Psychology* 36, no. 5 (2015): 507–523.

Warren, Alan. *Singapore, 1942: Britain's Greatest Defeat.* Hambledon and London, 2002.

Weight, Richard. *Patriots: National Identity in Britain 1940–2000.* Macmillan, 2002.

Welch, David. *Nazi Propaganda: The Power and the Limitations.* Barnes and Noble Books, 1983.

Welsh, Frank. *Australia: A New History of the Great Southern Land.* Overlook Press, 2006.

Wong, Leonard et al. *Why They Fight: Combat Motivation in the Iraq War.* Strategic Studies Institute, 2003.

Younger, R.M. *Australia and the Australians: A New Concise History.* Humanities Press, 1970.

Zilincik, Samuel. "The Role of Emotions in Military Strategy." *Texas National Security Review* 5, no. 2 Spring 2022): 11–25.

INDEX

air force(s), 118, 129–30, 182
Alexander, Harold, 162
anti-tank weapons; in Europe, 156, 158, 162; in Malaya, 92, 100; in North Africa, 126, 130–31, 135
anxiety, 164, 166, 169, 105
Anzio, 163, 175–76, 179–84, 196
Arab, 16, 72, 105, 154
armor, 79, 95, 96, 155
Arya Samaj, 44, 49
Auchinleck, Claude, 78, 84–85, 93, 95, 103, 120

Bennett, Gordon, 141
Blamey, Thomas, 151, 153, 156–58
Blitz, 71, 101
Boer War, 34, 36, 57–58
Bose, Subhas Chandra, 88, 128
boycott, 6, 39, 42–45.
  See also noncooperation
Brooke-Popham, Robert, 118
Burma, 69, 95, 120, 121, 128, 166

Canada, 18, 34, 56, 97
censorship, 20, 116; Australian reports, 102–4, 106–7, 152–53; British reports, 72, 72n20, 77–78, 80–85, 170, 175, 178, 183; Indian Reports, 152–53, 164, 166–68.
  See also morale
Chesterton, G.K., 34
Churchill, Winston, 82, 87, 117, 120, 150
Clark, Mark, 162, 163, 175, 176

cohesion, xvi, xvii, 13–14, 22–24, 195–96, 199; Australian, 101, 108, 137–38; British, 71, 75–76, 81, 86, 125; Indian, 90, 91–92, 97, 129.
  See also esprit de corps
colonialism, 34, 39, 43, 120
confidence, 2, 79, 104, 105, 107, 183; in leadership, 84, 86, 138; in unit, xvi, 81
constructivism, 2, 3, 4–5, 7–8, 198
Cripps, Stafford, 163–164, 168
Curtin, John, 106, 138

democracy, xvii, 67, 120, 137–38, 140, 178, 184, 193, 195; in Australian identity, 104–105, 108, 137–38, 140, 152–54; in British identity, 70–73, 87, 121, 178; in Indian identity, 87, 98–99, 129, 163
desertion, 20, 84–85, 161, 194
discipline, xvi, 14–16, 50, 179–84, 194, 197–98; Australian, 57, 100–102, 107–108, 140–44, 156, 160–62; British, 74–76, 79–80, 82–84, 86, 123–25, 179–84; definition of, xv, 20–21; Indian, 91–96, 130–36, 170–71, 174
doctrine, 17, 37, 52, 69, 119, 196

effectiveness, xi, 1, 8, 10, 24, 195–6, 199, 219
Egypt, 68, 92, 100, 136, 150–51, 154; defense of, 74, 82, 84, 96, 107
elites, 5–6, 8, 197, 199

emotion, xvi, 80, 122, 127, 184, 197
empire, xxii, 29, 61, 70, 87, 165; in Australian identity, 55–56, 97–99, 105–108, 137–40; in British identity, 33–36, 70; in Indian identity, 47–48, 87–88, 127–29, 136
eager, 81, 143, 153
esprit de corps, xvi, 80, 91, 123–25, 136, 170. *See also* cohesion
equipment, xiv, 16–17, 20, 67–69, 120, 149, 198; Australian, 151, 153–154, 156, 157; British, 73–76, 82; German, 86, 94, 108
ethnicity, xxi, xxiii, 2, 4, 7–8; in the military, 50, 89–91, 124, 129, 134, 139, 199; in national identity, 31–33, 49

Fadden, Arthur, 150
faith, 104, 105,
famine, 33, 166, 168
fascism, 40, 71, 128–29, 153, 165, 177–78
fear, 3, 6, 11, 134, 139–40, 178
feminism, 3, 198

Gandhi, Mahatma, 35, 40–43, 45–46, 96, 128, 164–65
German; military equipment, 69, 79, 86, 101, 107, 157–60; and Soviet Union (*see* Soviet Union); threat to Australia, 97–99, 106, 117, 137–40, 151; threat to Great Britain, 69–73, 86, 120–22, 127, 176–77, 183; threat to India, 87–89, 128, 163
goals of war, xiv, xviii, xxi–xxii, 1–3, 10–11, 192; Australian, 103–104, 152–154; British, 71–72, 86, 120, 176–78; and elites, 4, 8, 9, 15; Indian, 89–90, 163, 175; and national identity, 1–3, 21–22, 184
Greece, 68, 71, 73, 100, 103, 139; Australians in, 149–162

Greek, 151, 153, 154; military, 154, 155, 158
guerrilla warfare, 127, 143, 144

Hindu Mahasabha, 45, 46
Hitler, Adolf, 70–72, 82, 128
home, 10, 35
homesick, 104, 177,
honor, 13, 70, 86, 99, 150, 152; of regiment, xvi, xviii, 124, 134, 184, 195

ideology, xviii, xix, xxi, 5, 13, 71
Indian National Army, 128, 132,
Indian National Congress, 87, 88–89, 96, 127–28, 164–65
initiative, 14–15, 119, 174, 181–84, 194, 196–97; Australian, 100–102, 107–108, 140, 142–144, 155–59, 161; British, 74–76, 79, 8–86, 124–27, 181–84; definition of, xv–xvi, 21; Indian, 51, 91–92, 95–96, 129, 174
Ireland, 30, 32, 55
Italy, 68–70, 99, 117, 139, 162–184, 196

Japan; in Australian identity, 56, 57; and INA, 88, 128, 131–32; military equipment, 123, 124–26, 131, 135, 143–44, 196; military tactics, 119, 123, 126–27, 135–36, 142, 143–44; threat to Australia, 69, 72, 98–99, 104–106, 117, 119, 137–40; threat to Great Britain, 72, 117, 120–22; threat to India, 88, 94–95, 127–29, 163–66
Jewish, 13, 72, 178
Jinnah, Muhammad, 165

leadership, 90, 157; changes in, 69, 86, 101, 108, 125, 127; in democracy, xviii, 24, 83, 150; small unit, 51, 75, 79–80, 92, 133, 160

legitimacy, xxii, 1, 3, 7, 10–11, 14
Libya, 67–68, 77, 81, 94, 120, 167; Germans in, 74, 80, 158; Italians in, 98, 100, 15; racism towards, 154, 177
Lucas, John, 175–76, 182

Malayan, 116, 119, 121–22, 129,
manpower, xiv, xv, 200
martial classes, 50–52, 89
Mechili, 74–75, 91, 95, 100
Menzies, Robert, 54, 98, 150
Monte Cassino, 162–63, 169–75, 179, 196
morale, 14–16, 72n20, 194, 197; Australian, 100–102, 104, 106–108, 140, 144, 152; British, 74, 77–86, 122–23, 125–27, 182–84; definition of, xv, 20–21; Indian, 90, 93–94, 96, 131–36, 166–68, 171–74
Moreshead, Leslie, 101,
Muslim League, 47–49, 88, 127, 164–65
myth, xiv, xix–xxiii, 28–30, 192–94, 197–98, 200; in Australian identity, 53, 99, 107–108, 152–53; in British identity, 38, 70, 73, 76, 80–82, 86; definition of, 2–3, 4, 21–22; in Indian identity, 49, 165
myth-symbol complex, xxi, 2, 7

Napoleon, 35, 70, 82
narrative, 4–5, 21, 22, 90, 200
nation, xxi, 2, 5–8, 21–22
national identity; Australian, 105, 138–40, 153; British, 70–73, 76, 86, 122, 127, 176; changes to, 8–9, 18, 197; definition of, xxi, 1, 21–22; and emotion, 2, 11–13, 197; Indian, 129, 165; and race, 2, 72, 197; and war goals, xxi–xxii, 3, 9, 194, 199–200

nationalism, xvii, 9, 15, 89–90, 130, 197
Naujawan Bharag Sabha, 41
Navy, 31, 59, 83, 117, 158, 160
Nehru, Jawaharlal, 41, 96, 128, 164
New Zealand, 17, 56, 150–51, 153, 163, 171–74
noncooperation, 39, 40, 43–44, 48, 165

Operation Compass, 100
Operation Crusader, 78–79, 81, 93
Operation Matador, 118, 123, 130

Pakistan, 39, 47, 90, 164
panic, 106, 123, 125, 132–133, 135, 156
patriotism, 10, 34, 36, 80, 87, 193
poll. *See* public opinion
power: creation of, 27, 196–97; military, 30–31, 57, 118, 125, 152; political, 47–48, 56, 88, 98, 121, 164; and threat, xviii, 57, 117, 140
pride, 2–3, 11, 13, 15; in unit, 23–24, 29, 167, 183–84
prisoners of war (POW), 15, 102, 107, 116, 161; Australian, 141, 161; democracies and, xvii, 24; German, 75, 94, 155, 161, 182; Indian, 88, 107, 128, 132, 136, 167
propaganda, 3, 10, 24, 132, 138–39, 200; British, 78, 87, 117, 122, 128; German, 78, 102
public opinion, 22, 98, 121, 152, 176

Quit India, 96, 164–66

race, 54–56, 72, 90, 105, 137, 178
racism, 18, 72, 92, 139, 177, 197
rationalist, 3–4, 198
realism, xvii, xix
recruitment, 29, 36–37, 50–51, 59, 89, 132
religion, 29, 67, 91, 129, 167, 199

replacement, 23, 29, 36
respect, 83, 93, 99, 102, 139
retreat, xv, 20–21, 166, 196; in Europe, 156–59; in North Africa, 74–75, 77, 81–85, 92, 95, 100–103; in Malaya, 125–27, 131, 133–36, 144
rotation system. *See* replacement
Russia, 57, 198; war with Germany, 70, 82, 88, 96, 152, 167, 183. *See also* Soviet Union

Savige, Stanley, 159–160
Scottish, 30, 37, 80, 178
Sikh, 131, 175; identity, 40, 46, 49, 167–68, 195; military units, 90, 93, 96, 109, 134
skill. *See* training
Slim River, 125–26, 134–35
Socialism, 40, 61, 87
Soviet Union, 40, 70, 72, 89, 103. *See also* Russia
sorry, 166, 176–77
South Africa, 40, 70, 72, 89, 103
Stewart, Ian, 119, 122, 123–25, 127, 141, 196
strategy, xv, xviii, 4, 50, 68, 101; Singapore, 53, 59, 97, 117–18, 137
surrender, xvii–xviii, 83–85, 98, 116, 140; of Australian soldiers, 100, 161; of British soldiers, 75, 78, 83; of Indian soldiers, 92, 95; of Singapore, 116, 140
symbols, 71, 86, 153, 192–94; definition of, xx–xxi, 2, 6, 22; elite use of, 3, 6, 7–9, 197; and emotion, xxii, 2, 4, 11–13

tactics, 50, 60, 69. *See also* initiative
Thailand, 118, 123
threat, 6–7, 12, 24, 197; to Australia, 137–140, 98–99, 137; in discipline, xv, 15, 20, 133; to Great Britain, 70–71, 120–122, 185; to India, 94, 127, 164–165; motivating soldiers, 86, 88, 120–22, 133–37, 140, 164; and nationalism, xvii–xix, 29, 71, 122, 129, 191
Tobruk, 83–84, 158, 160; Australian units in, 99, 101–103, 107–108; British units in, 72, 75–80; Indian units in, 91–94
training, xv, xvi, 15–7, 69, 119, 196; Australian, 58–60, 100, 104, 107–108, 142; British, 36–38, 74, 79, 86, 124, 127; and cohesion, 23–24, 29; Indian, 50–51, 91, 97, 136
Turkey, 41, 47–49, 150, 161

United Nations, 96, 178
United States, 5–6, 8, 33, 72, 78, 120; military units, 162–63, 169, 179–83

volunteer, 21, 37, 50; in battle, 142, 150, 165, 194; for military service, 58–60, 87, 89–90, 106, 128, 139

Wavell, Archibald, 69, 74, 78, 103, 151, 158
weapons, xv, 60, 120, 131, 134–35, 156
Welsh, 30, 32–4
Winter Line, 162, 175
World War I, 10, 41, 43, 56, 78, 87–88; in identity, 35–6, 48–51, 58, 153

ABOUT THE AUTHOR

Kirstin Brathwaite is an associate professor of international relations at Michigan State University, where she teaches in James Madison College. She has published on ethnic conflict, combat effectiveness, urban warfare, and civil-military relations.